The Fabulous Contraptions of Jasper J. Pumpkinhead

by Brian Kesinger

Baby Tattoo Books
Los Angeles

Trade Edition ISBN: 978-1-61404-027-9
Limited Edition ISBN: 978-1-61404-115-3
Digital Edition ISBN: 978-1-61404-207-5

Design and Title Typography by Mark Berry

Cover Animation by Tina Nawrocki

10 9 8 7 6 5 4 3 2 1

Published by
Baby Tattoo Books®
Los Angeles

babytattoo.com
Find the artist online @briankesinger

Manufactured in China
Fabriqué en Chine

INTRODUCTION

Brian Kesinger has worked for two decades as a contributing artist on some of history's most popular animated motion pictures. Having spent countless hours creating make-believe worlds from his drafting table, he enjoys devoting his leisure time to filling sketchbooks with true-to-life studies of some of the amazing people, places and things he encounters while exploring and observing our wonderful planet.

This catalog is devoted to the inventive oeuvre of a remarkable individual who has not allowed geographical isolation or lack of significant social status to deter him from dreaming bigger than the rest of us. Never having traveled farther than the farm where he was born, he is living proof that one does not know where the next great genius might come from. The sketches assembled here provide technical renderings of his prototype innovations as works-in-progress. Perhaps this showcase of his ideas will establish his place in history as a contemporary da Vinci (but more fascinating due to his having a pumpkin head).

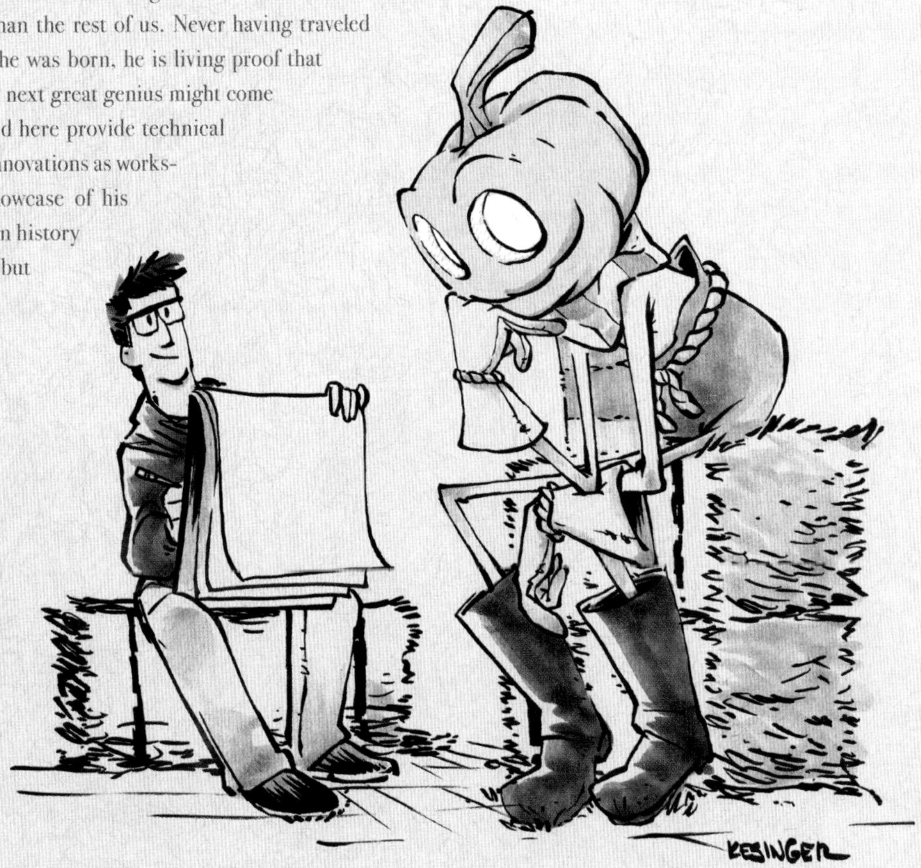

Brian & Jasper

PATENT 2044B:

"The Bucket-Kicker"

Upcycling is a great way to invent new things on a small budget.

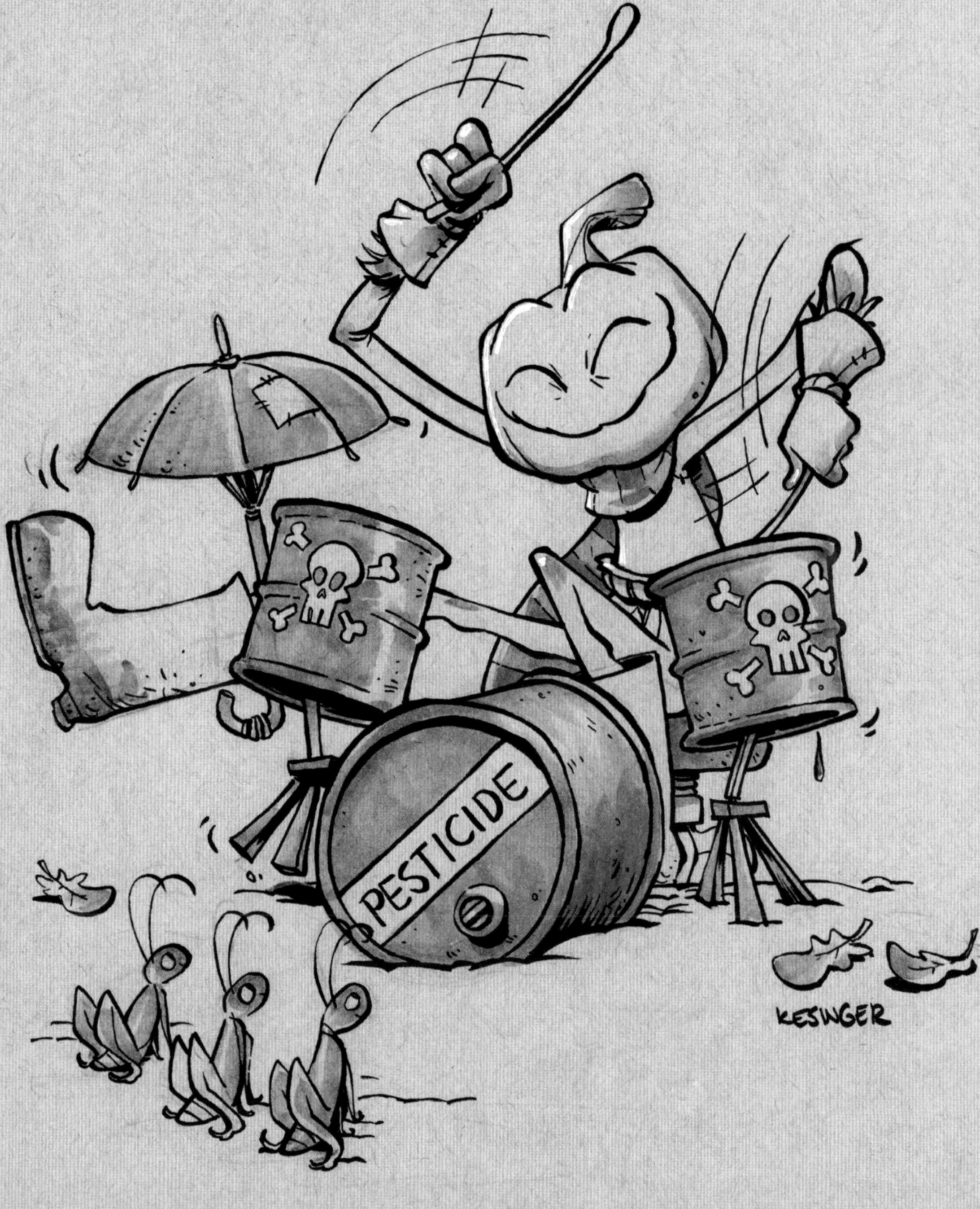
PESTICIDE
KESINGER

PATENT 140.55:

"The Swingamabob"

This contraption is the best way
to enjoy brisk autumn evenings.

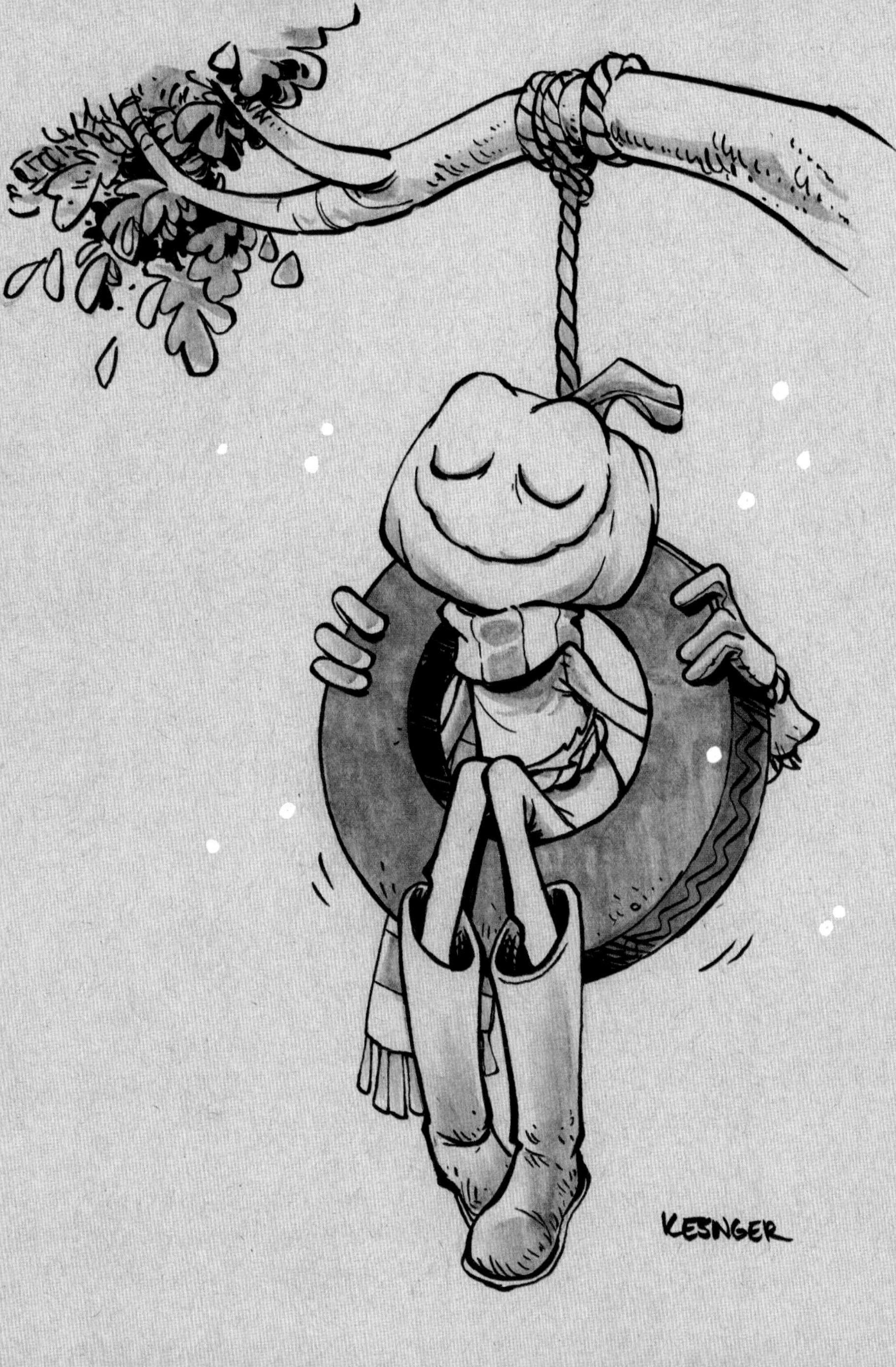
KESINGER

PATENT 13599-Q:

“The Handy-Dandy”

This lengthy device insures that marshmallows
are the only things that turn warm and gooey.

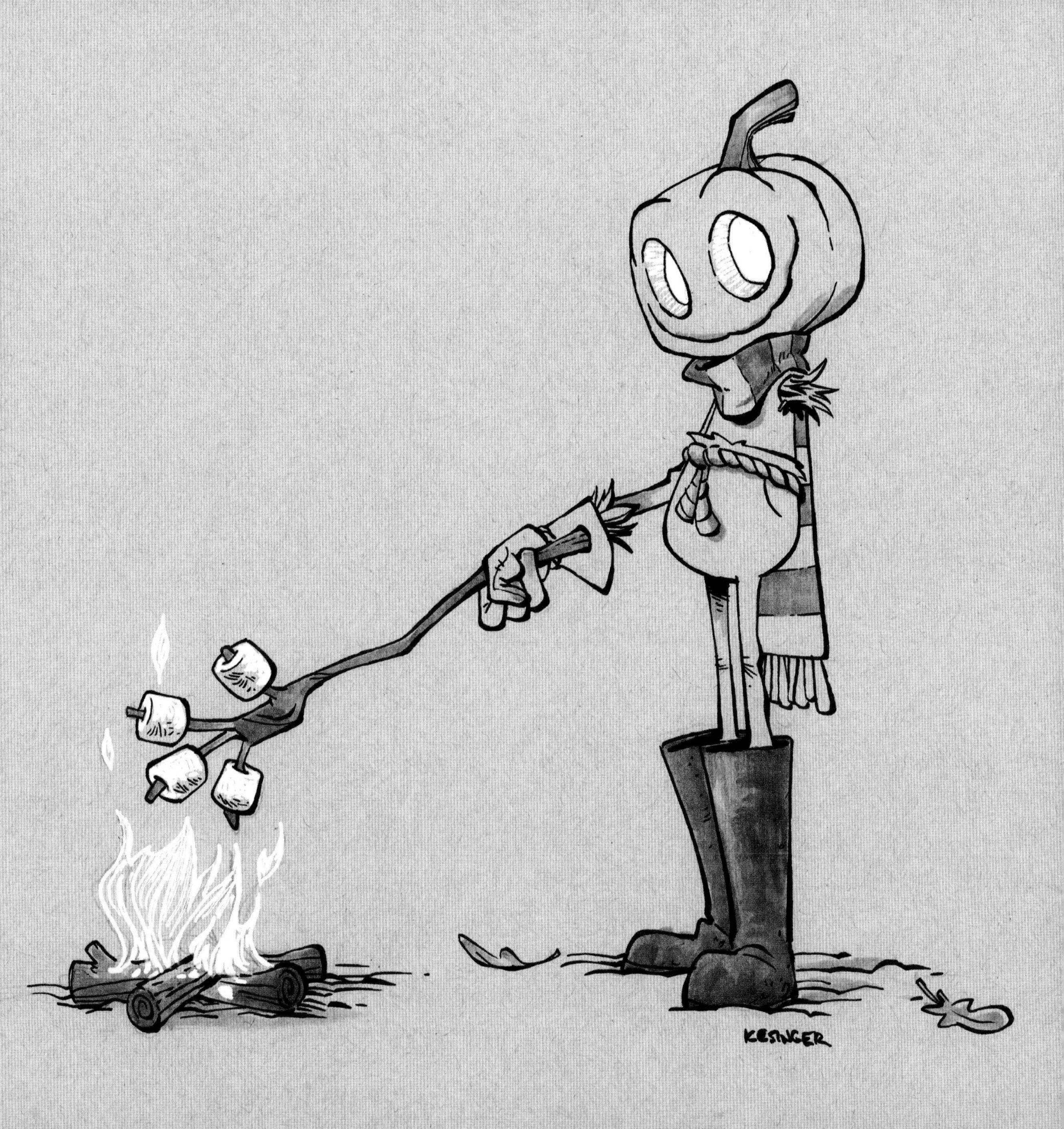
KESINGER

PATENT 55:

"The Eye-Pad"

**With it's amazing 5-inch chalkboard surface,
this will help you organize your thoughts one word at a time.
(Chalk and eraser not included)**

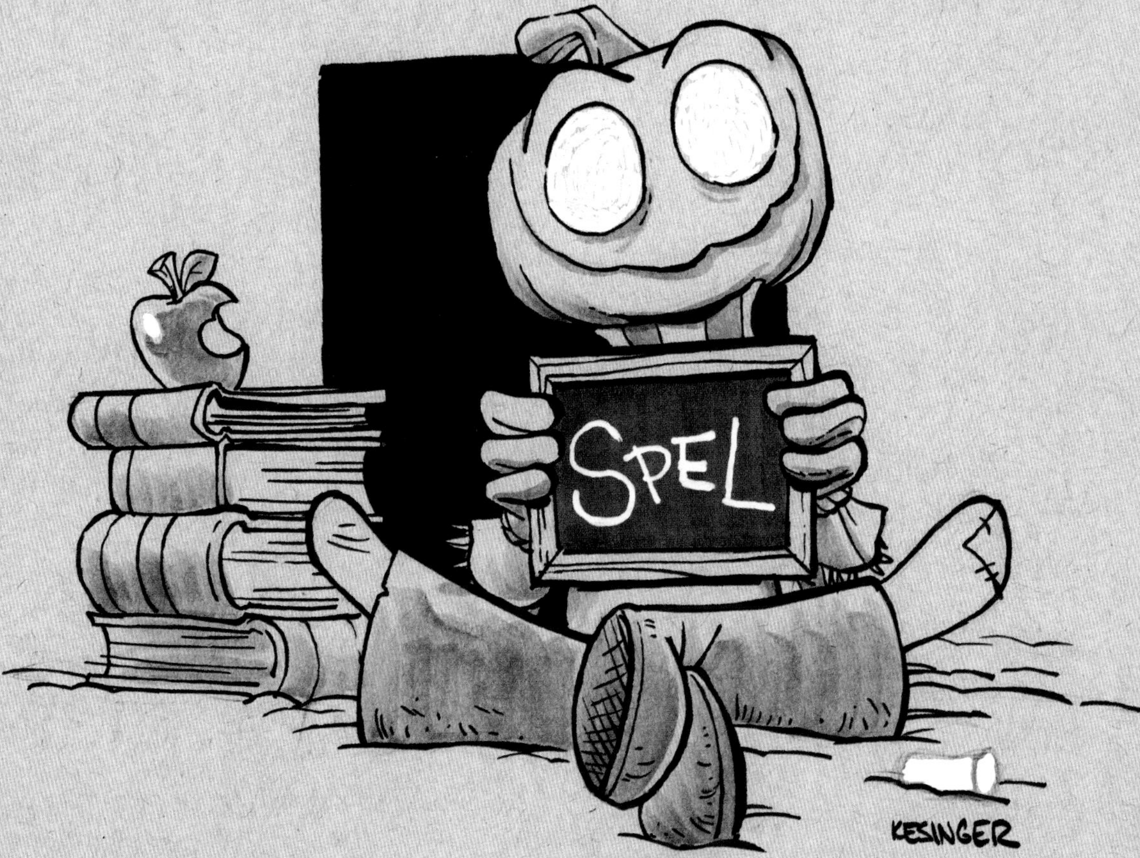
SPEL
KESINGER

PATENT 1313:

"The Feed-a-Fone"

Daily chores are breeze with this invention.
Leaving more time for building!

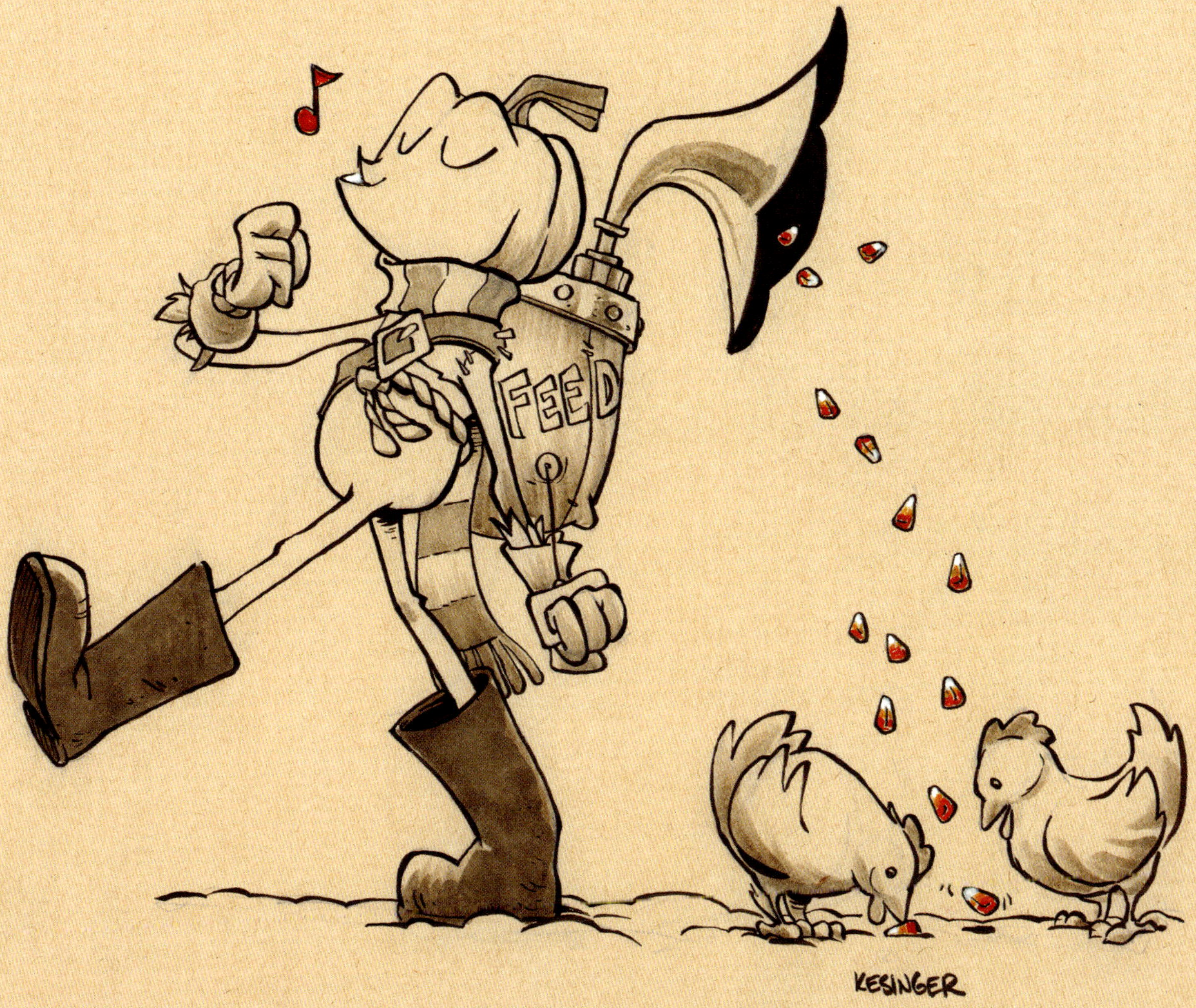
FEED
KESINGER

PATENT 140.55:

"The Pie-Detector"

Harnessing the nostril power of a twin-turbine engine, you'll never miss a freshly baked good again.

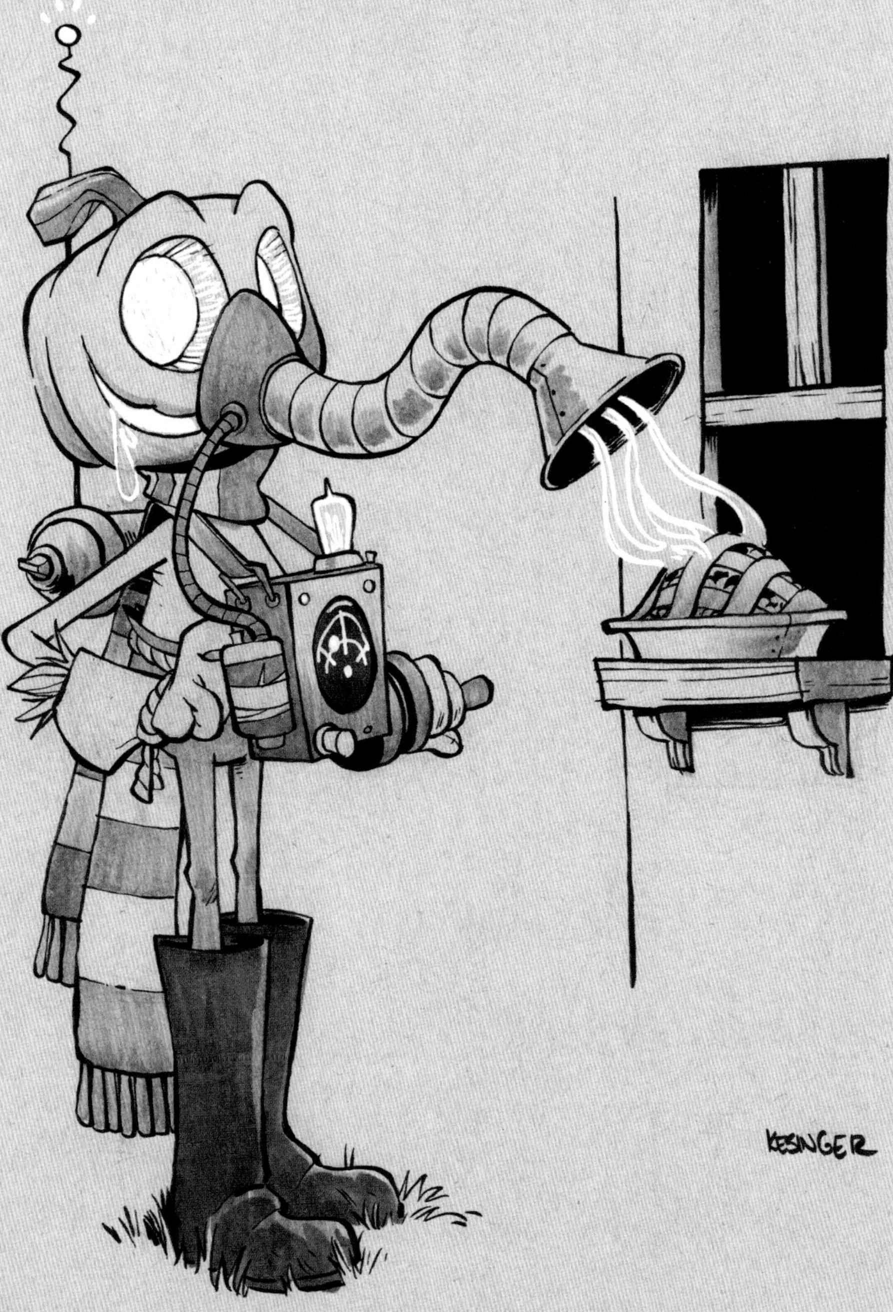
KESINGER

Protip:

Resting in-between projects is
the best way to dream up new ideas!

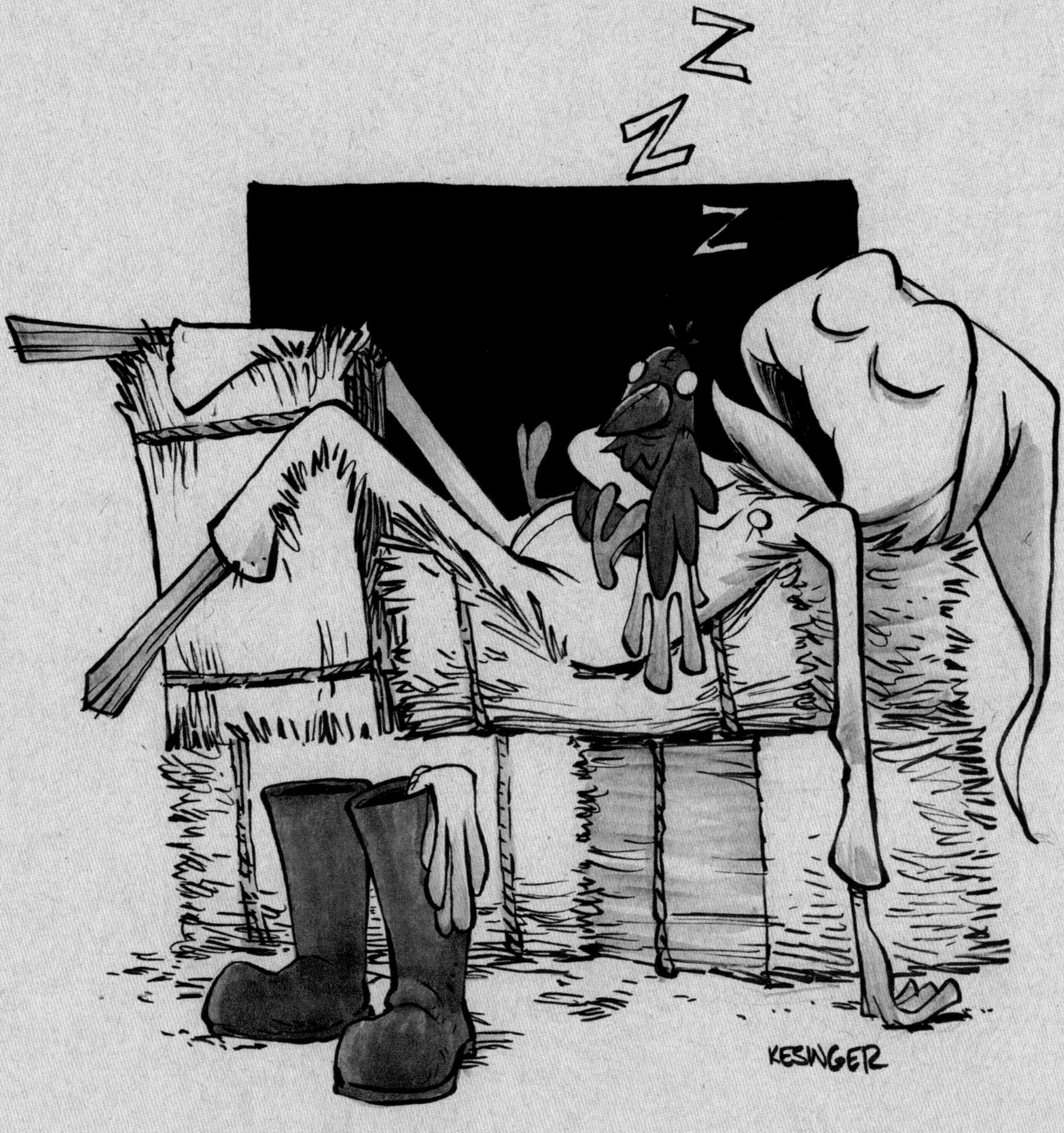

Z
Z
Z
KESINGER

PATENT 123456789:

"The Stellarscope"

This ocular invention will ensure a great view of the stars no matter the weather conditions.

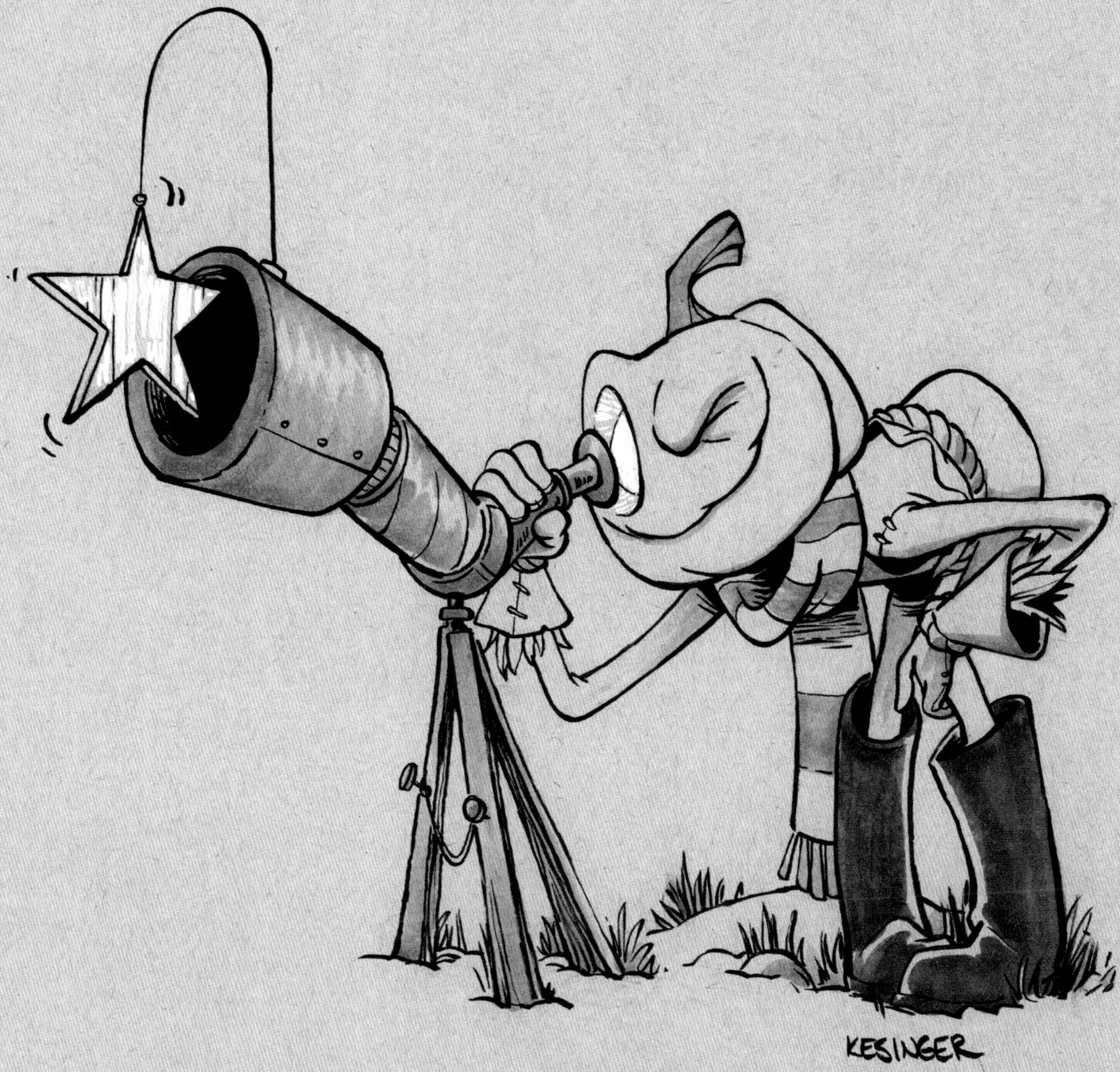
KESINGER

Protip:

When trying to incubate new ideas,
remember to use your head.

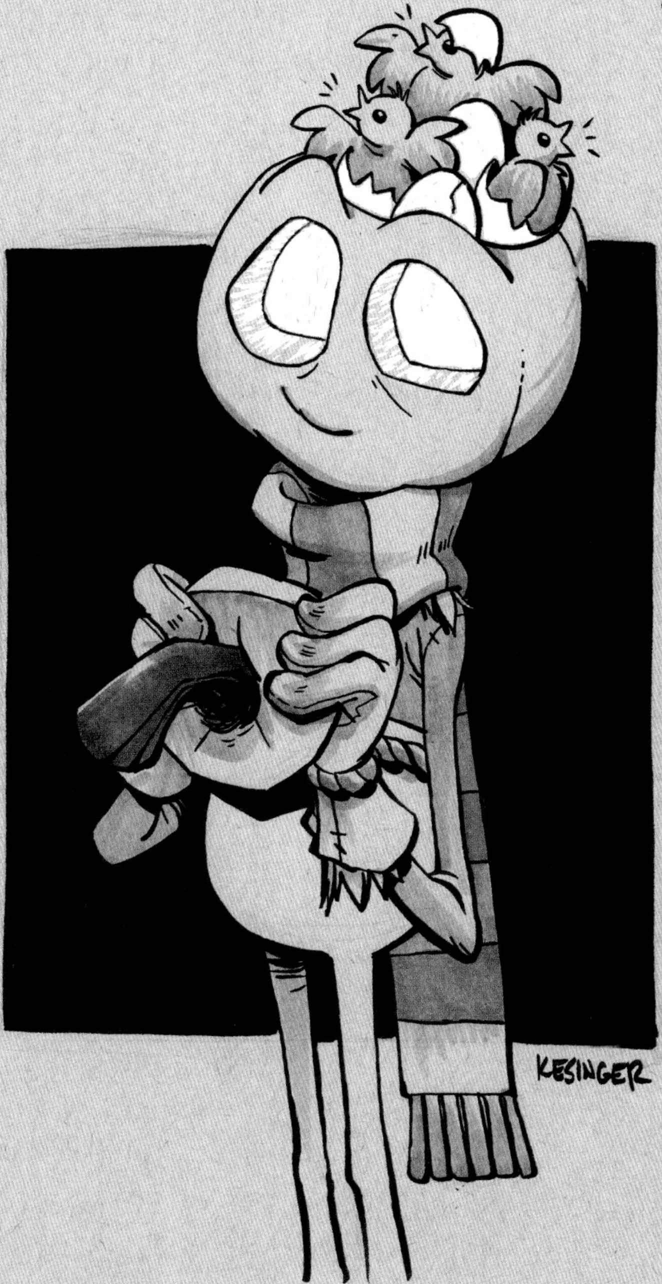
KESINGER

PATENT 529:

"Flood Pants"

Cut your gardening time in half
with these dapper dowsers.

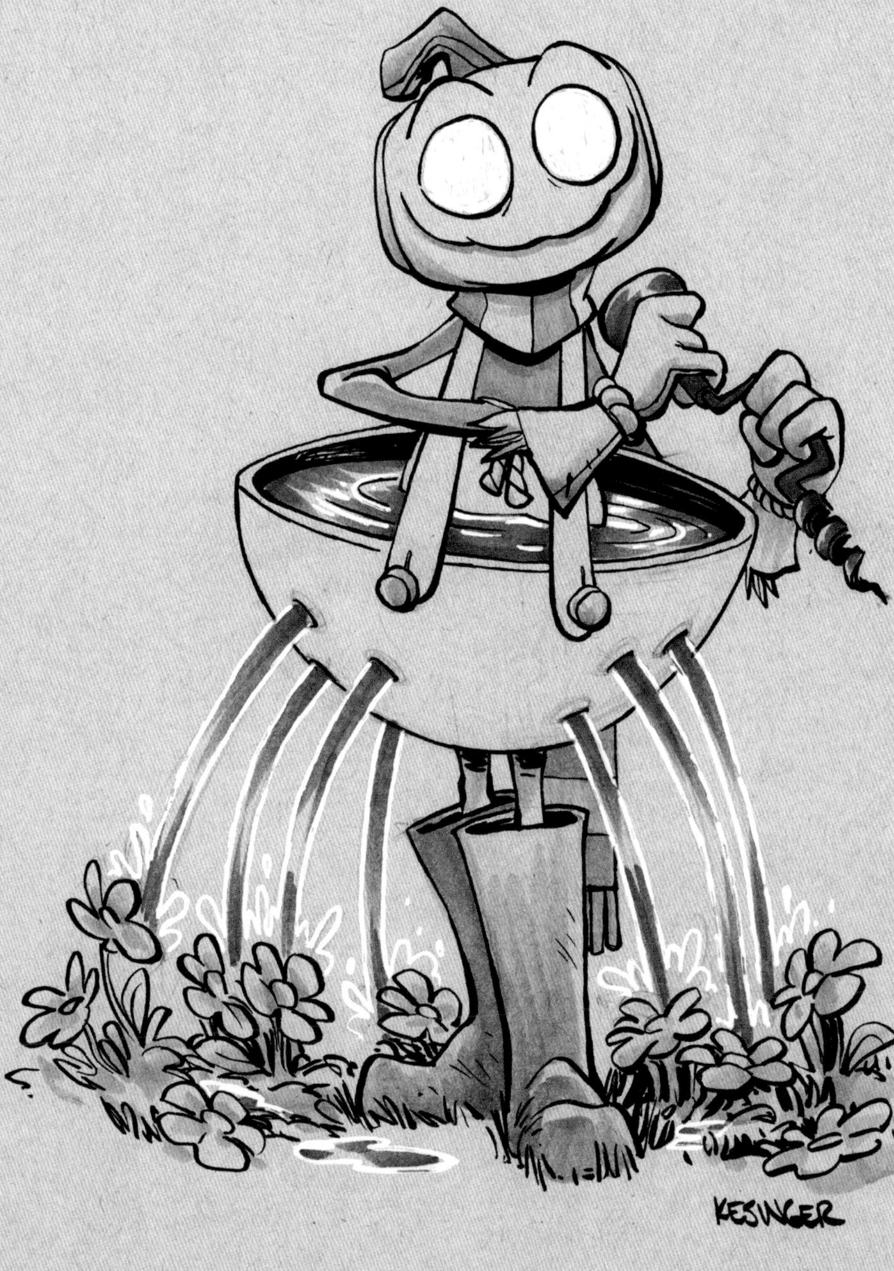
KESINGER

To-do list item #345:

Invent a less barbaric way
to decorate for Halloween.

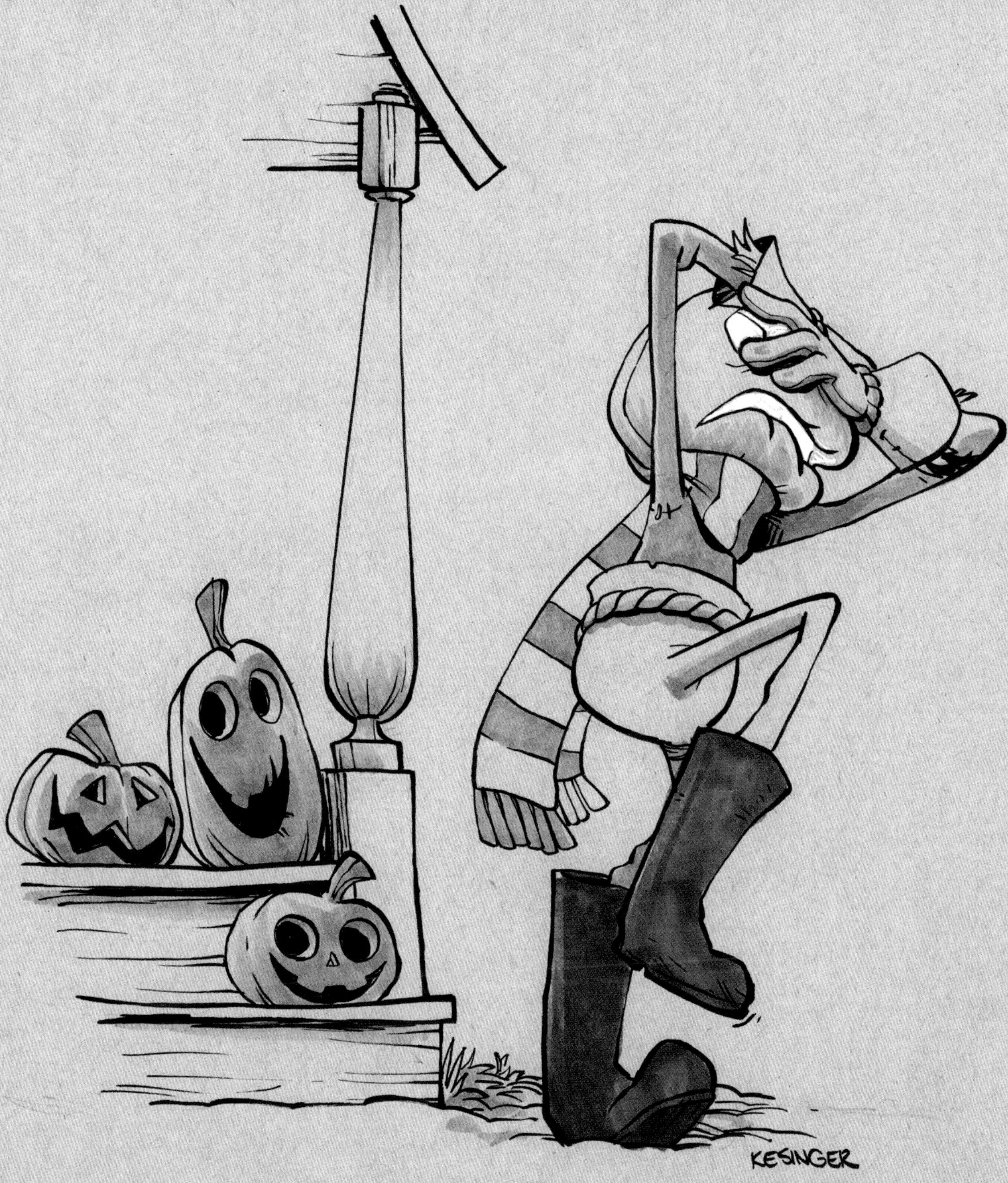
KESINGER

PATENT 12C:

"The Floaty-Fish"

Once inflated, this will ensure hours of fun and safe water play!

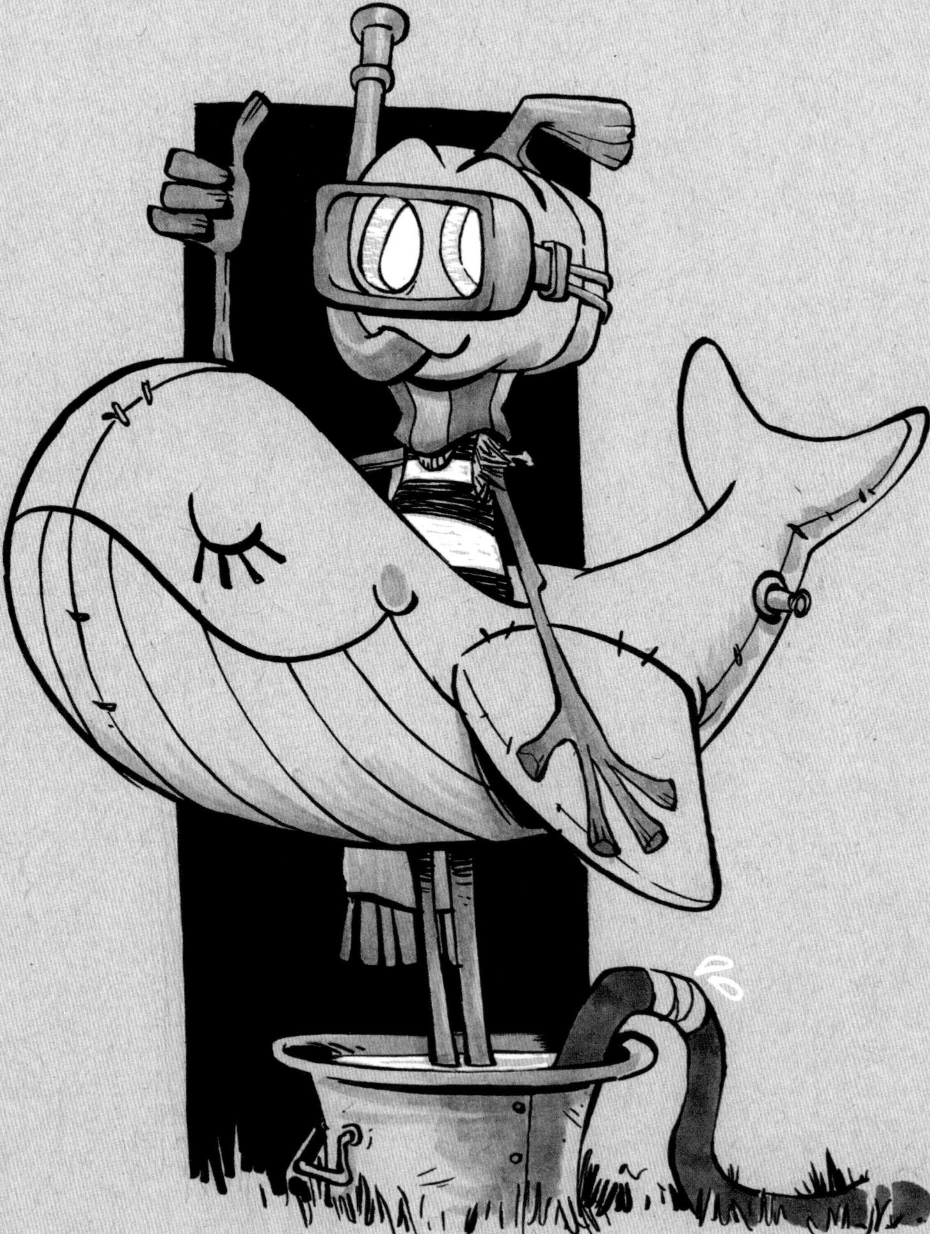
KESINGER

PATENT 6002:

"The Nighty-Knight"

Scaring crows is serious business
and one does not simply do so unprotected.

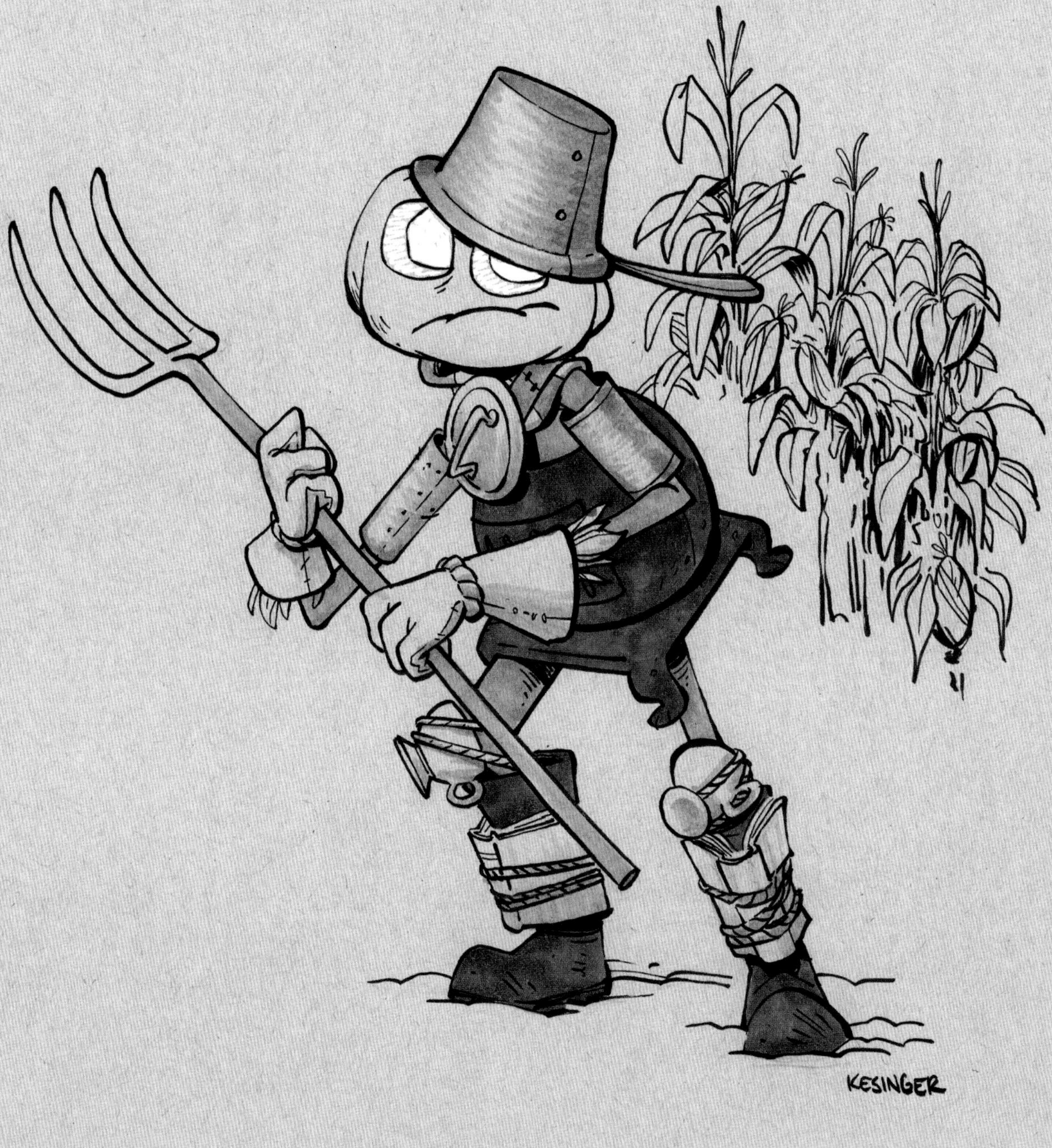
KESINGER

PATENT 0400:

"The Al-Arm Clock"

If you are cuckoo for new technology, this especially stylish wristwatch is for you!

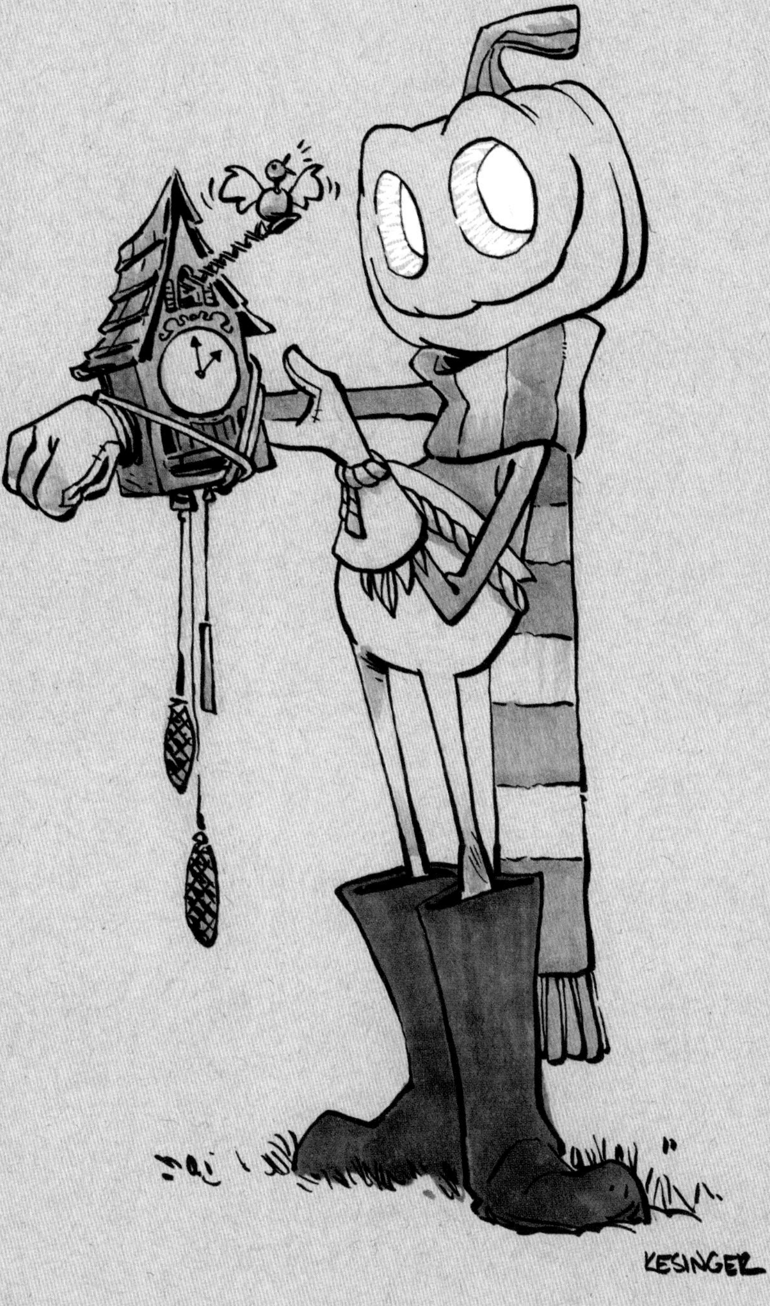
KESINGER

To-do list item #7892:

Invent a muscle suit
for lifting heavy objects.

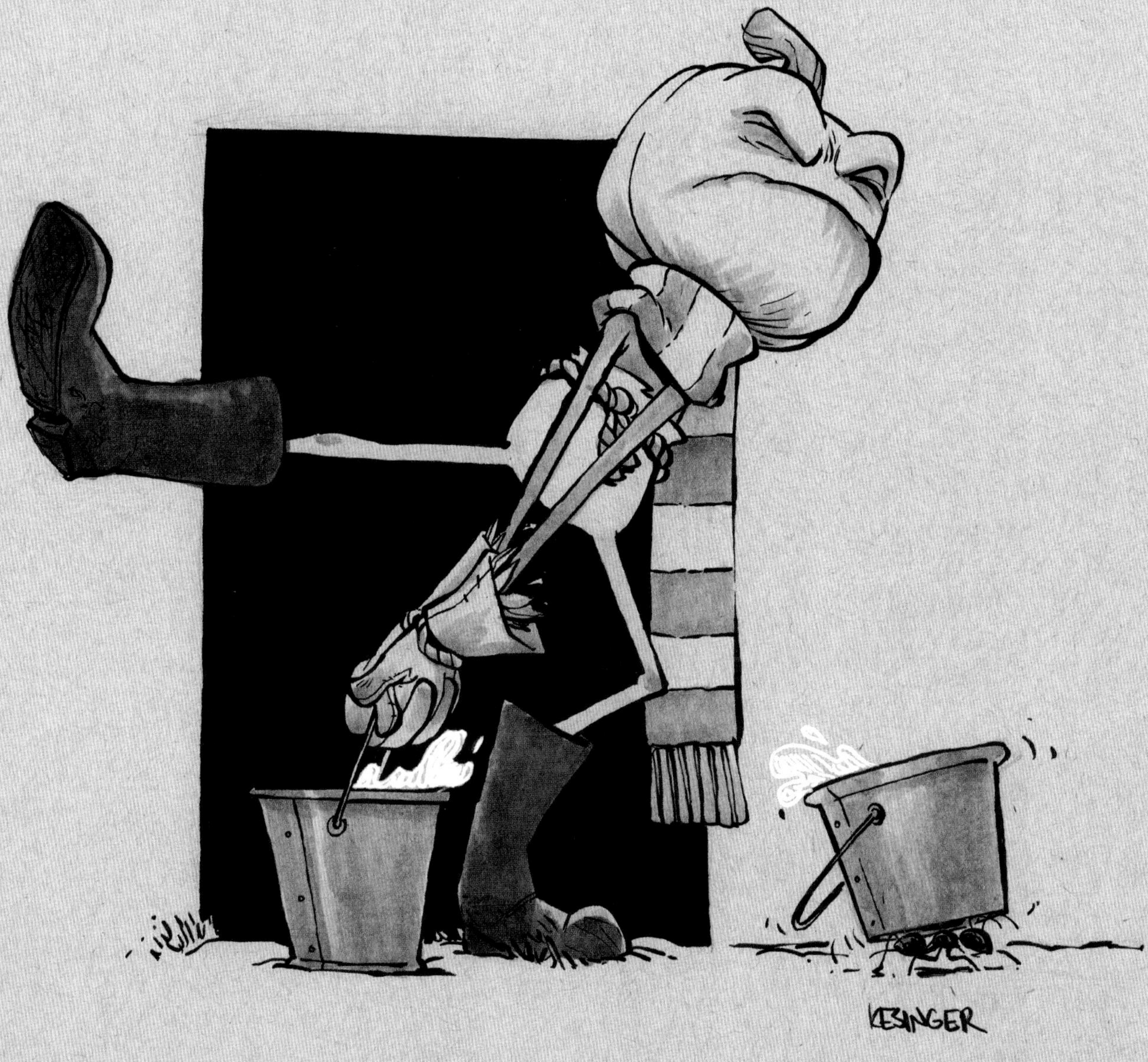
KESINGER

Protip:

A well lit workspace and a solid drafting table are required for thinking up the next big thing.

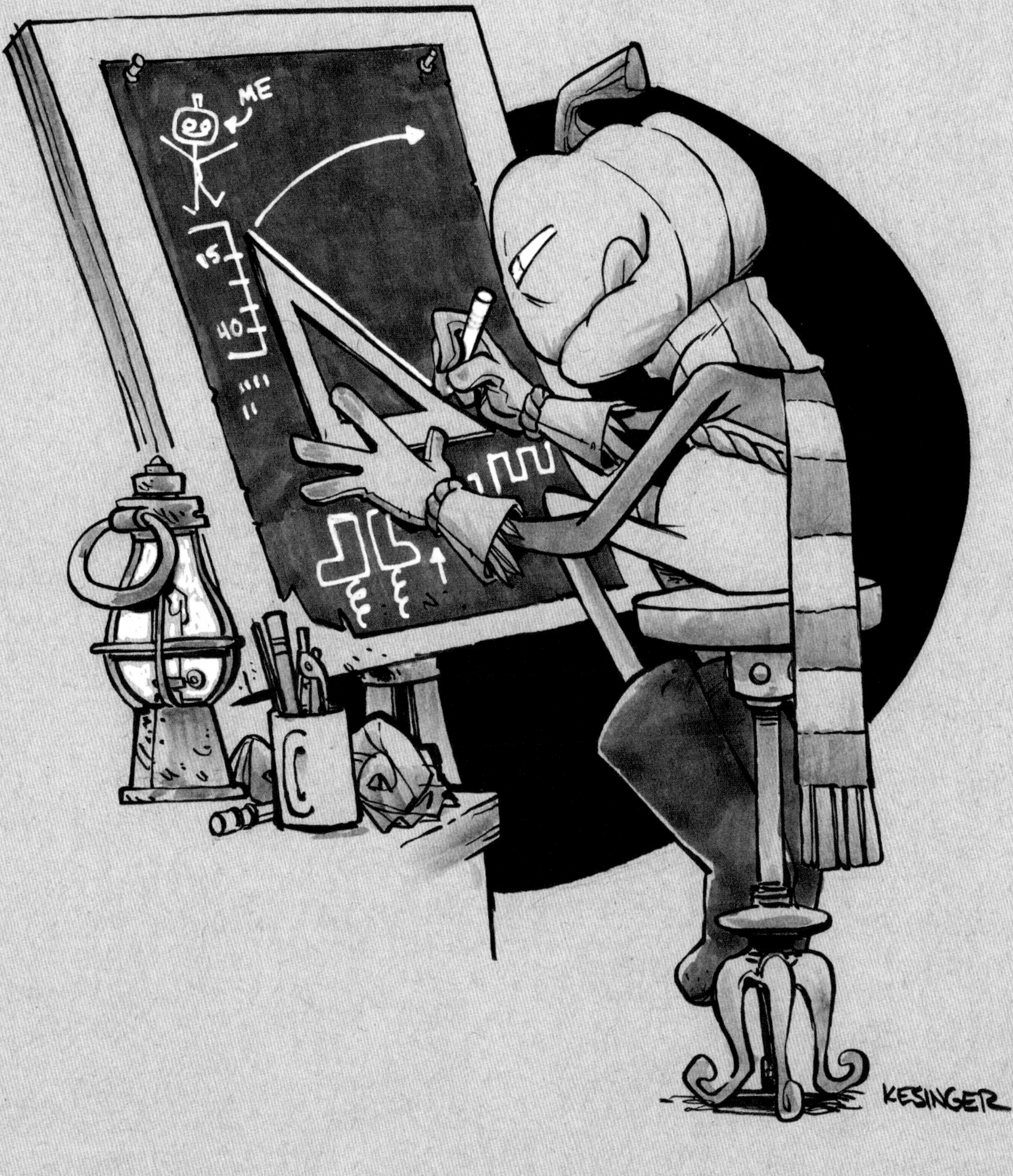
ME
40
KESINGER

To-do list item #1:

Invent a better way to collect honey!

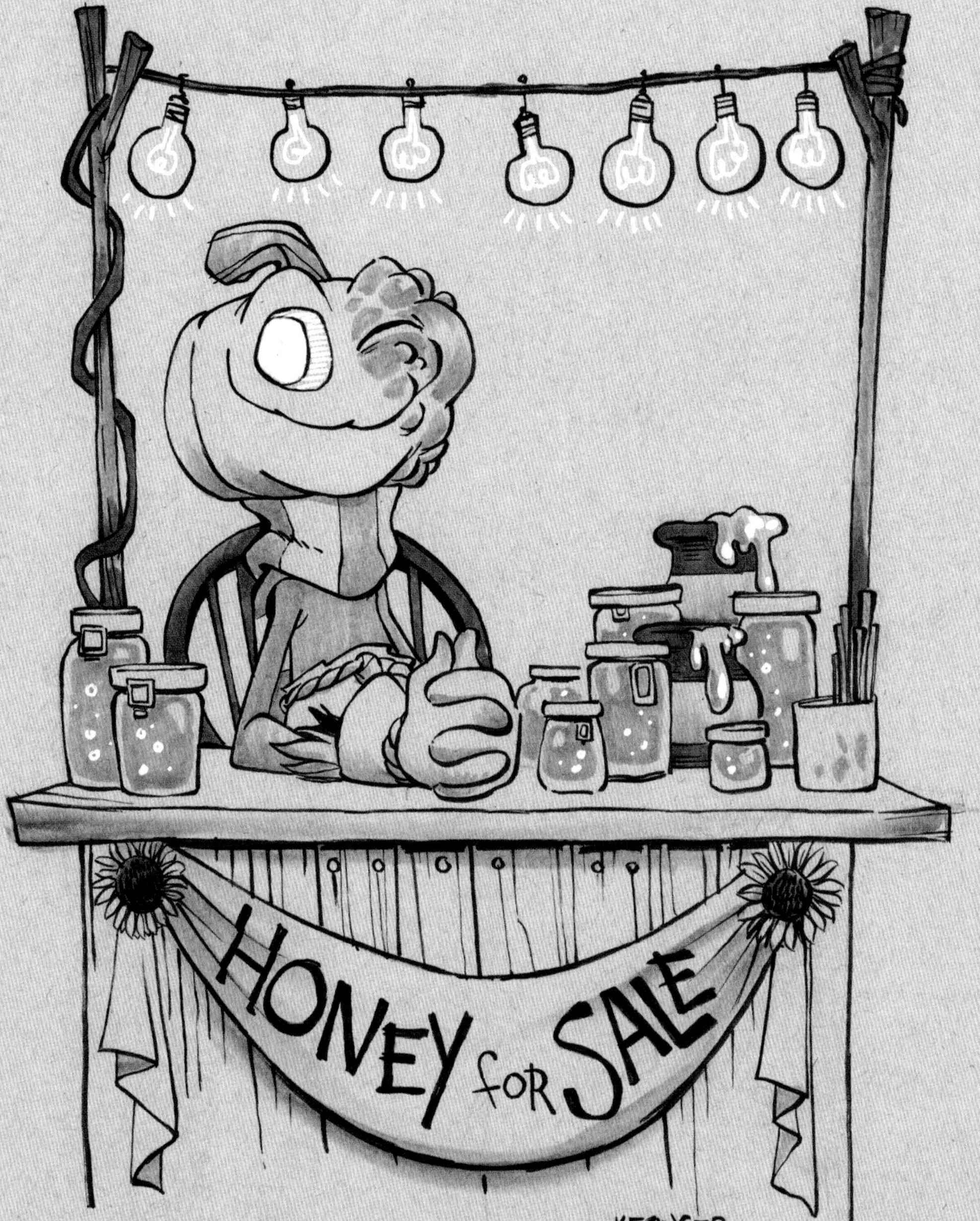
HONEY for SALE
KESINGER

PATENT 222.2:

"Twonoculars"

Using this optical enhancement will allow you to get an up-close look at all the amazing wildlife around the farm.

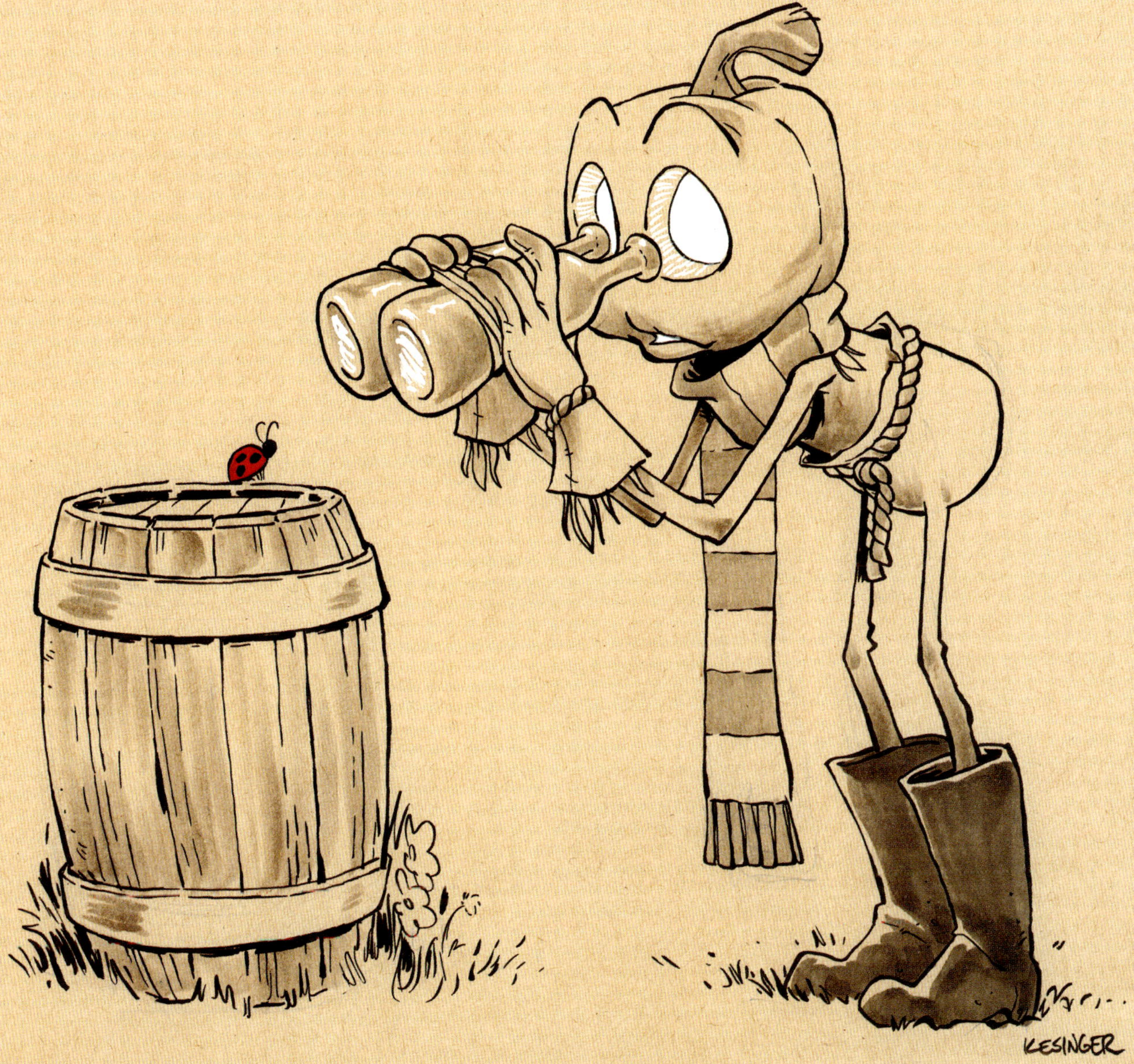
KESINGER

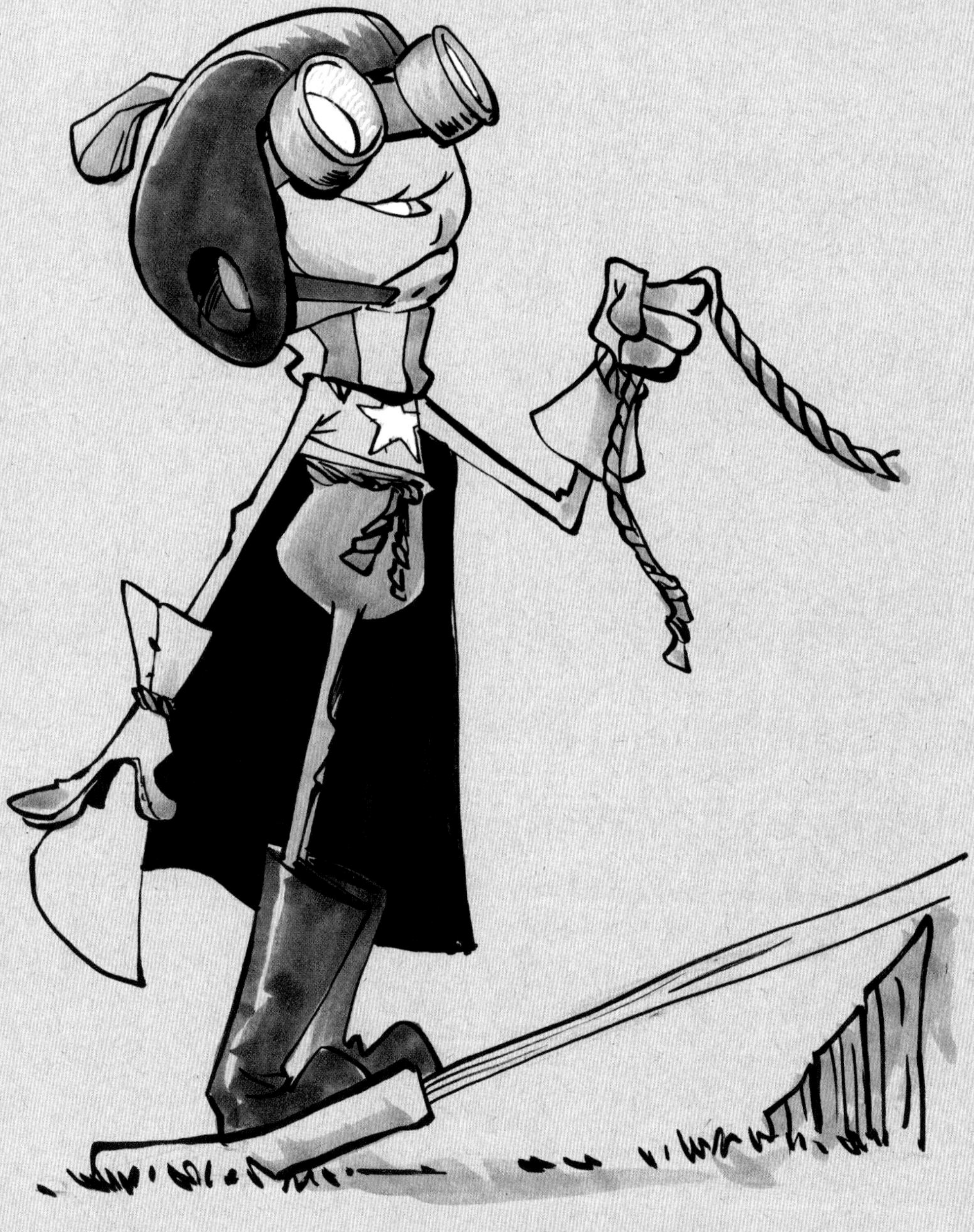

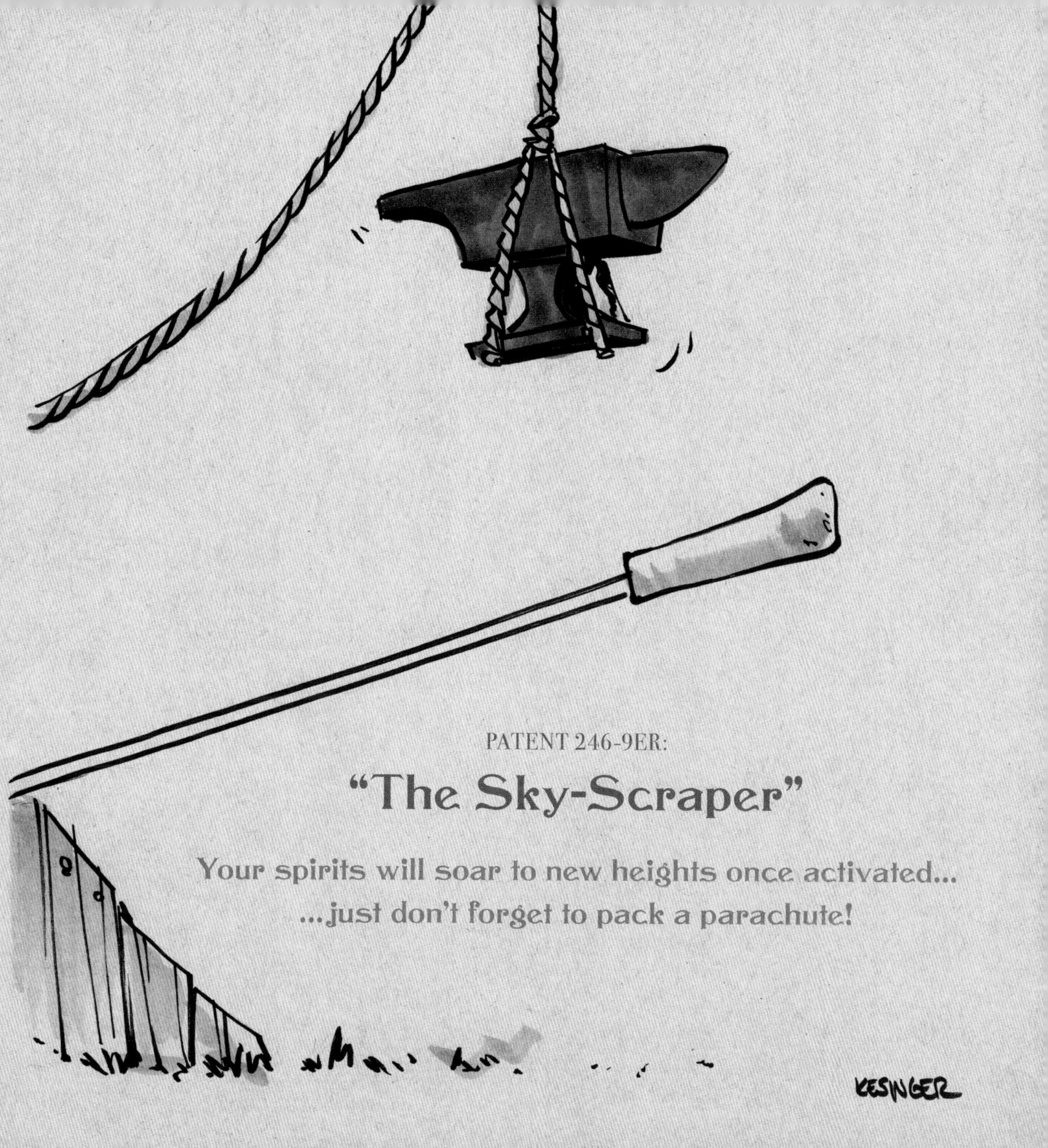
PATENT 246-9ER:
"The Sky-Scraper"
Your spirits will soar to new heights once activated...
...just don't forget to pack a parachute!
KESINGER

Protip:

Being an inventor may not make you rich
but sometimes you have to make sacrifices
to follow your dreams.

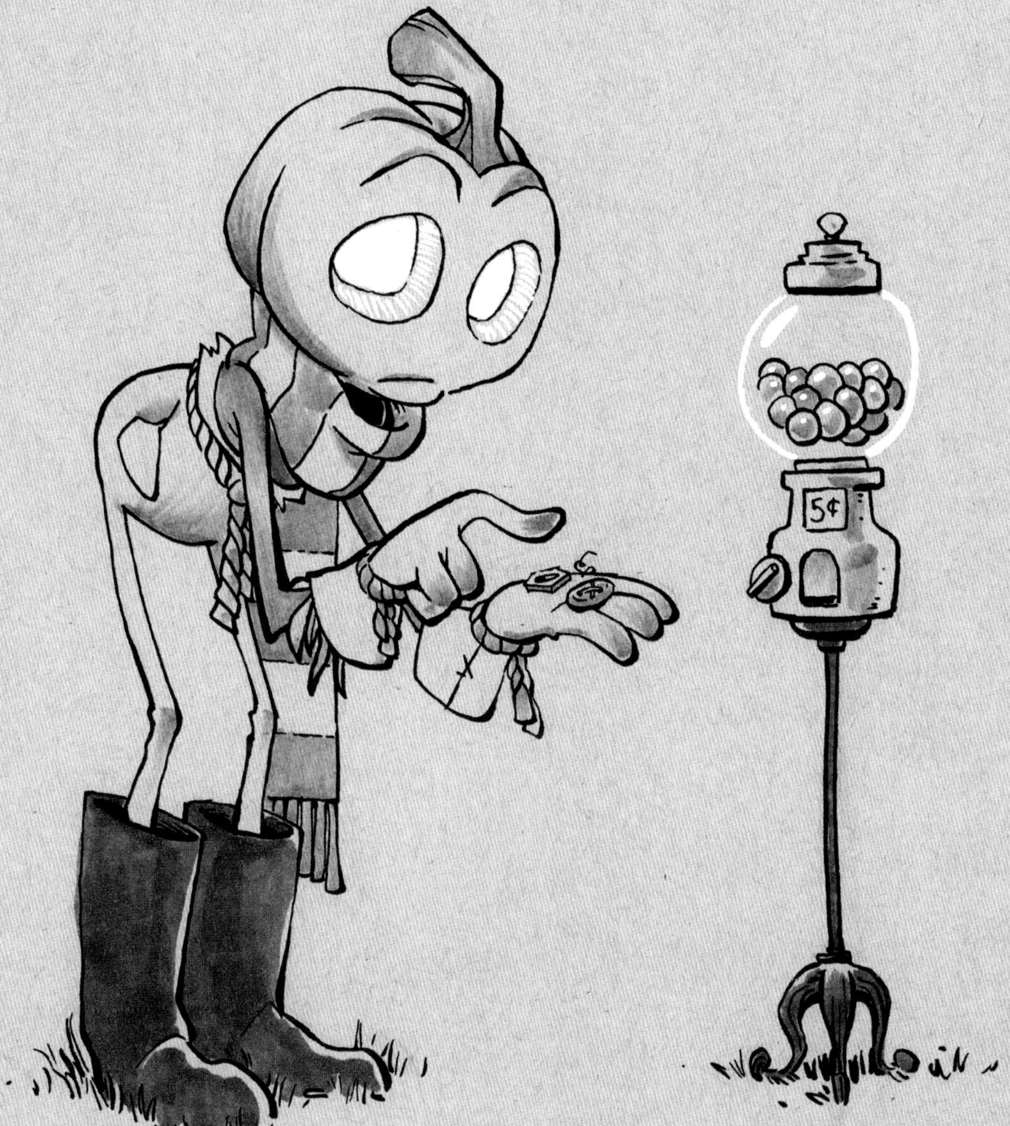
5¢
KESINGER

To-do list item #96:

Invent flame retardant clothing... and hay... and sticks (or stay away from the fire completely).

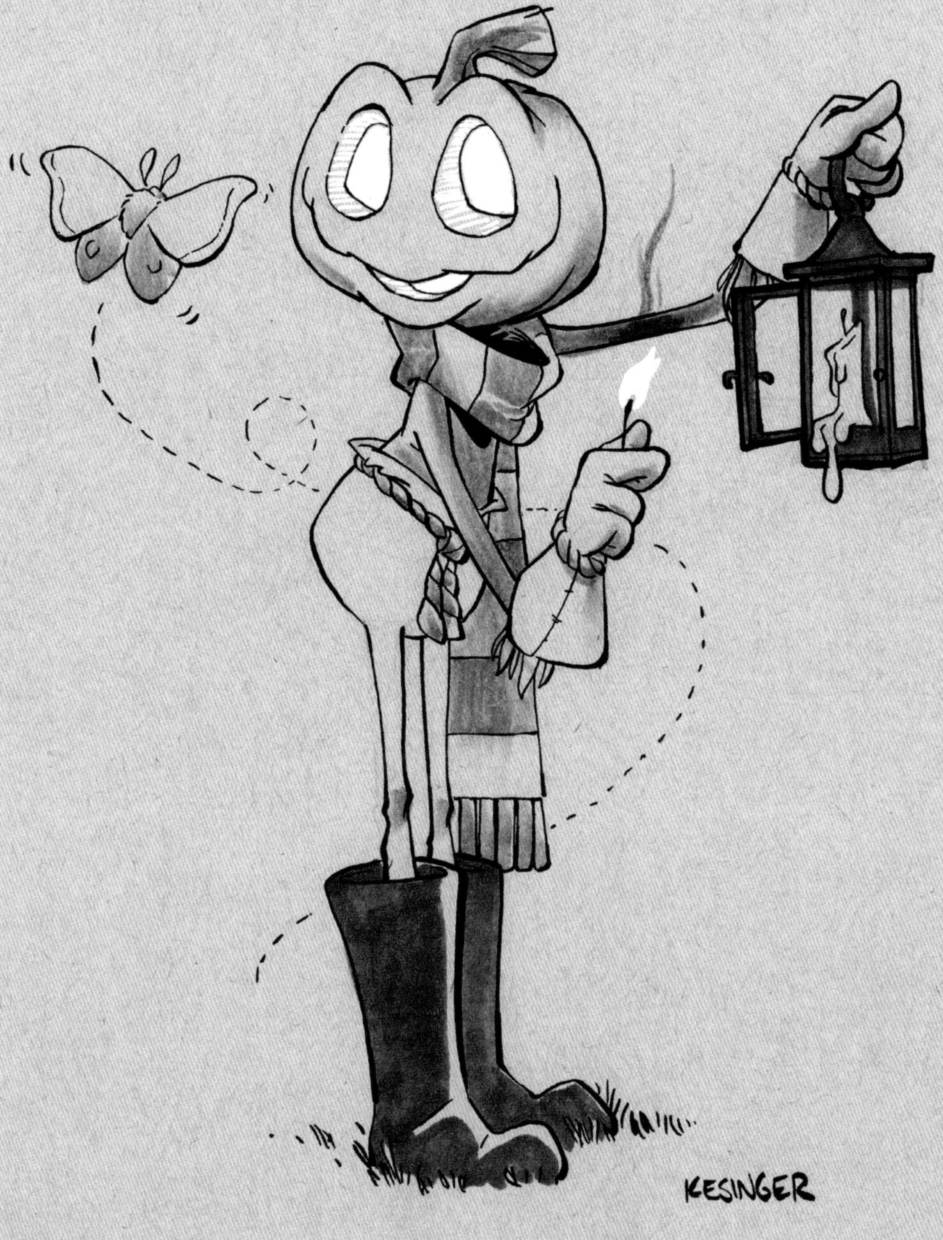
KESINGER

PATENT 3087.6B:

"The Hogwash"

Maintaining porcine hygiene has never been easier with this modern convenience.

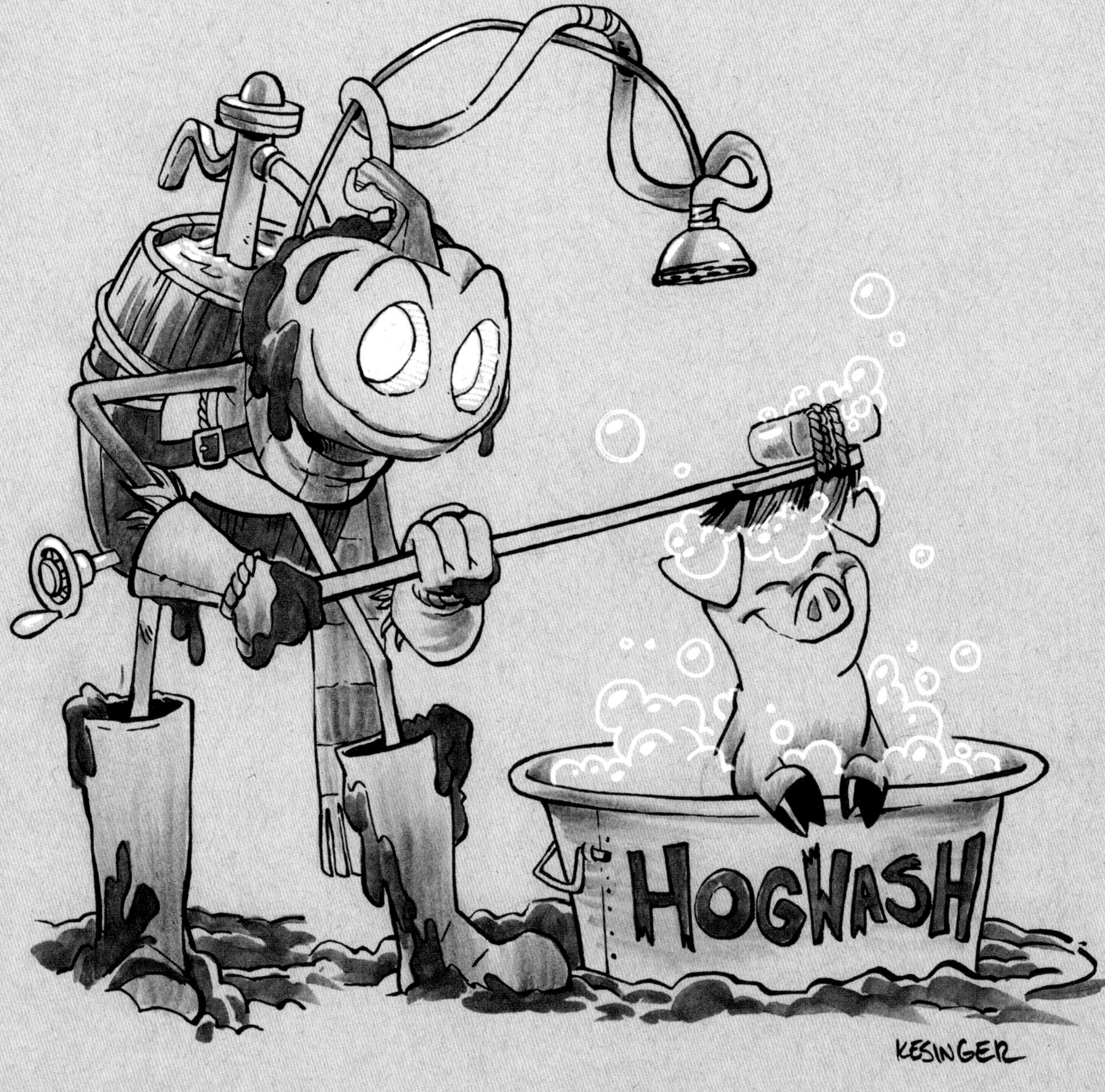
HOGWASH
KESINGER

PATENT 43.5-7:

"The Motorsaw"

This will cut your workload in half,
but be careful that's all it cuts.

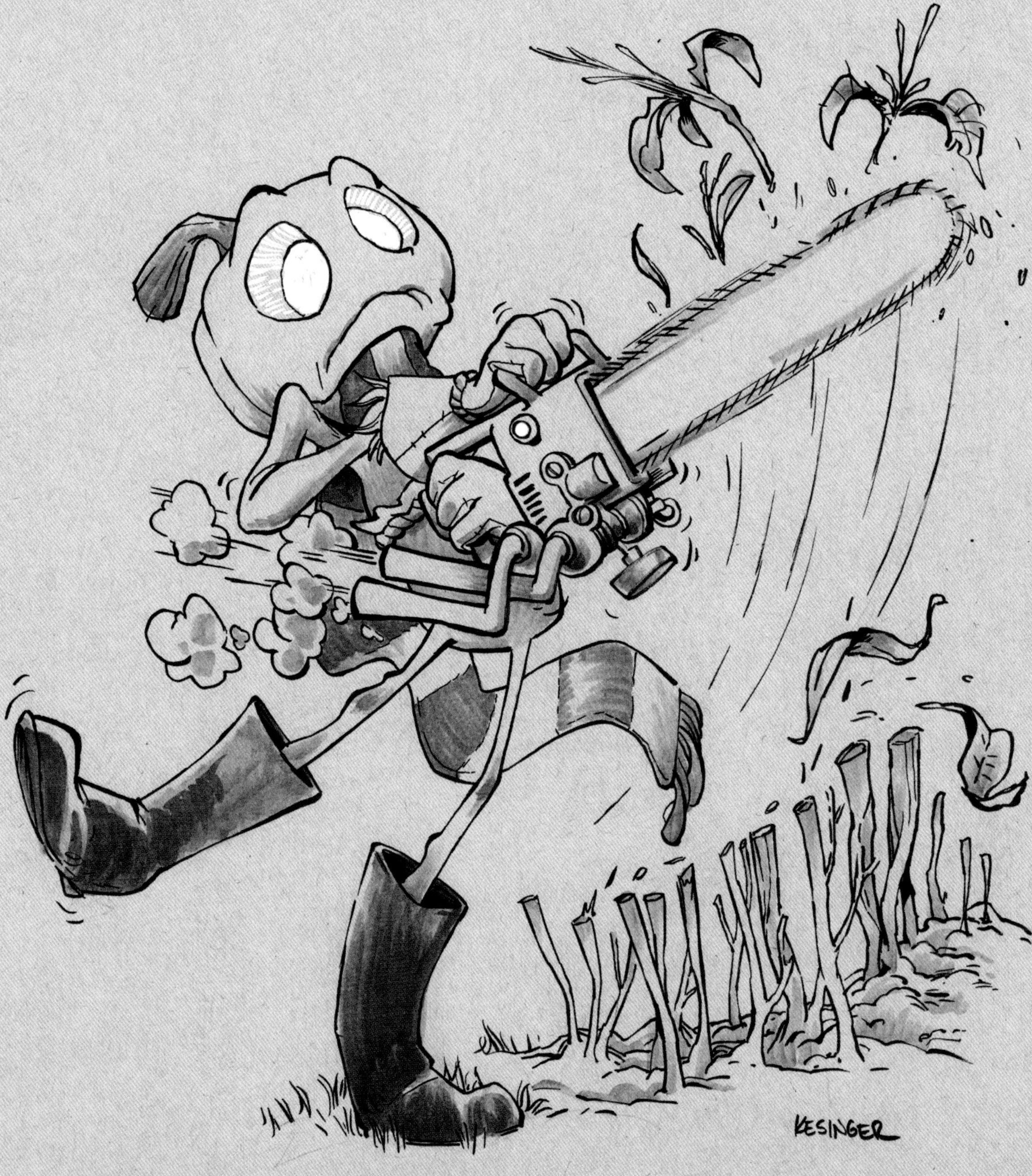
KESINGER

PATENT 322-D:

"The Clog-B-Gone"

Nothing backs up a sink like scarecrow straw, but that will be of no concern once this hose-matic contraption is deployed.

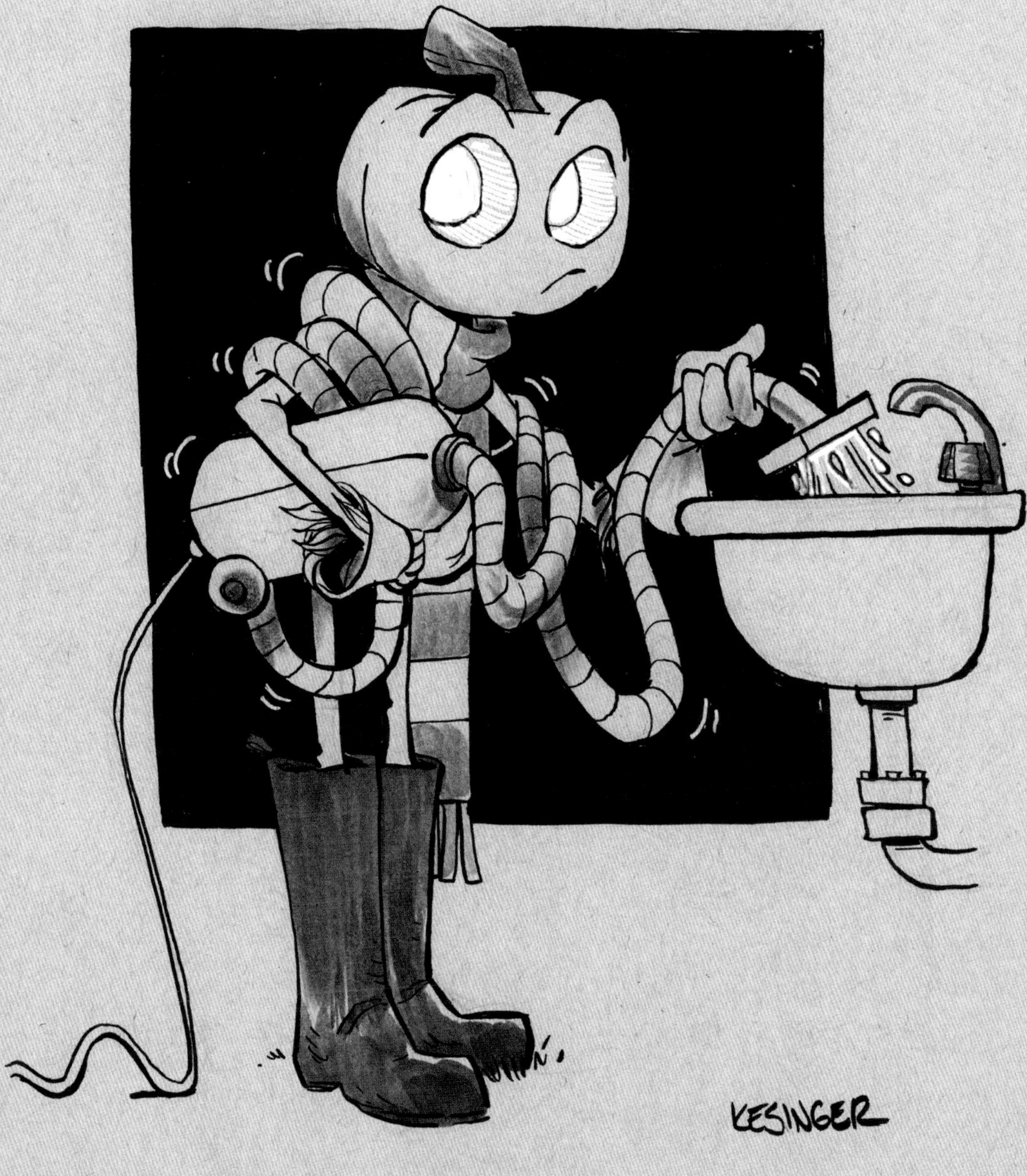
KESINGER

Protip:

Product testing is a stressful but necessary step in the process of inventing.

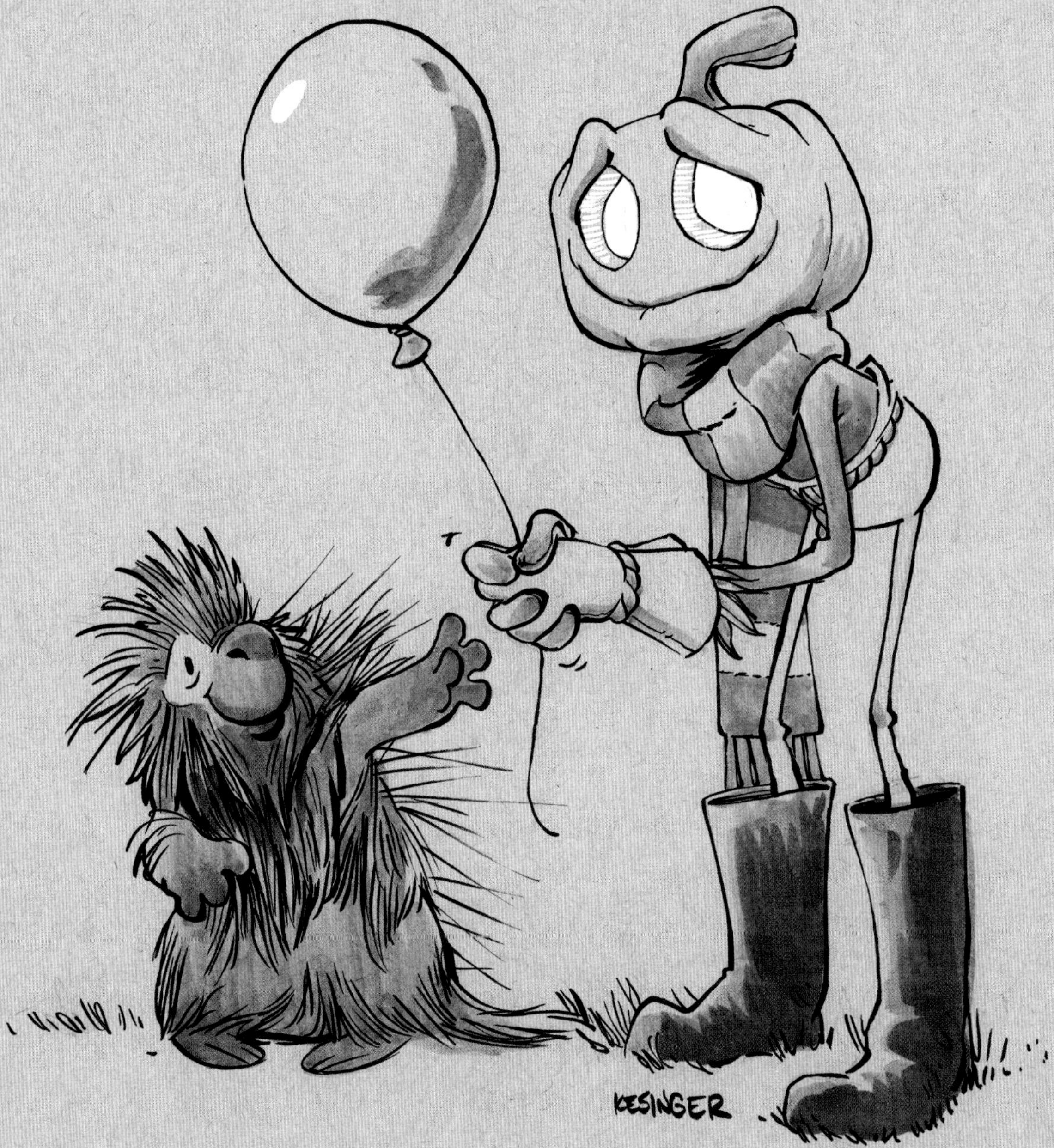
KESINGER

Protip:

Always take a moment to stretch
and stay limber in-between projects.

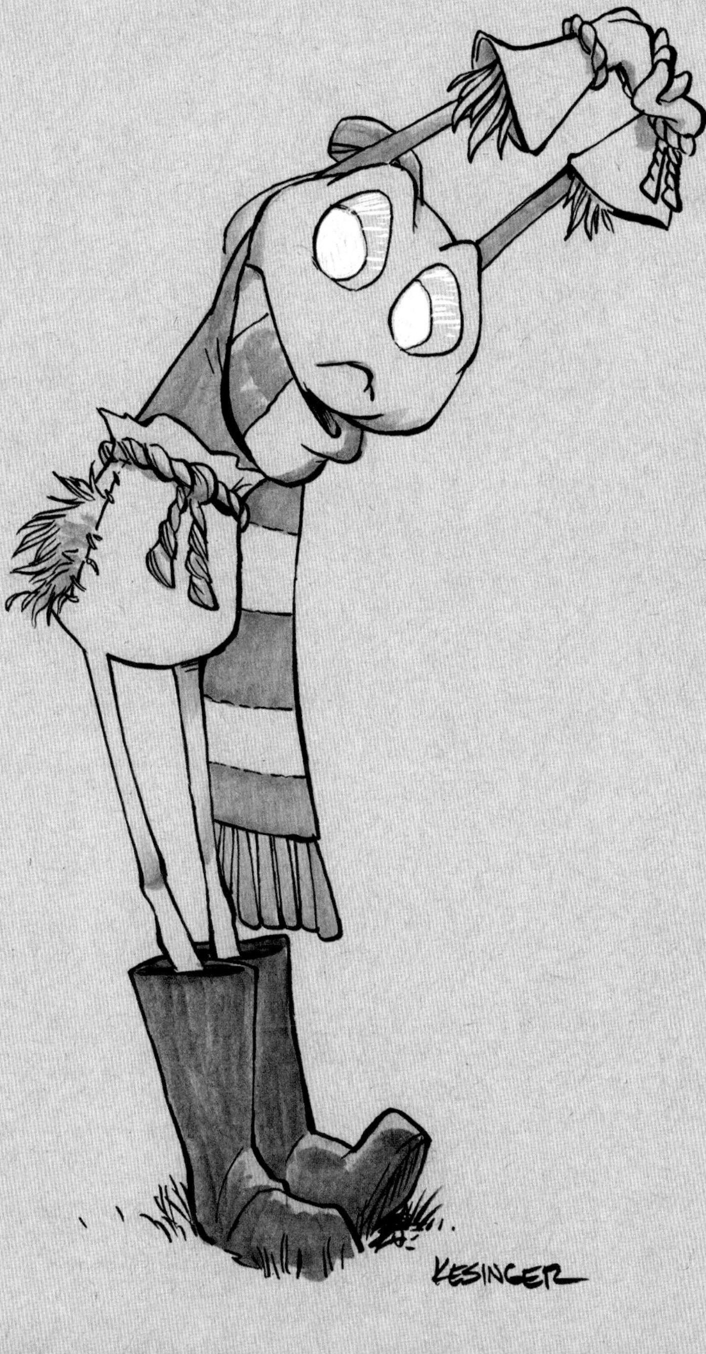
KESINGER

PATENT 0003.50:

"Boomstoppers"

These therapeutic apparatuses comfort barnyard creatures great and small during thunderstorms.

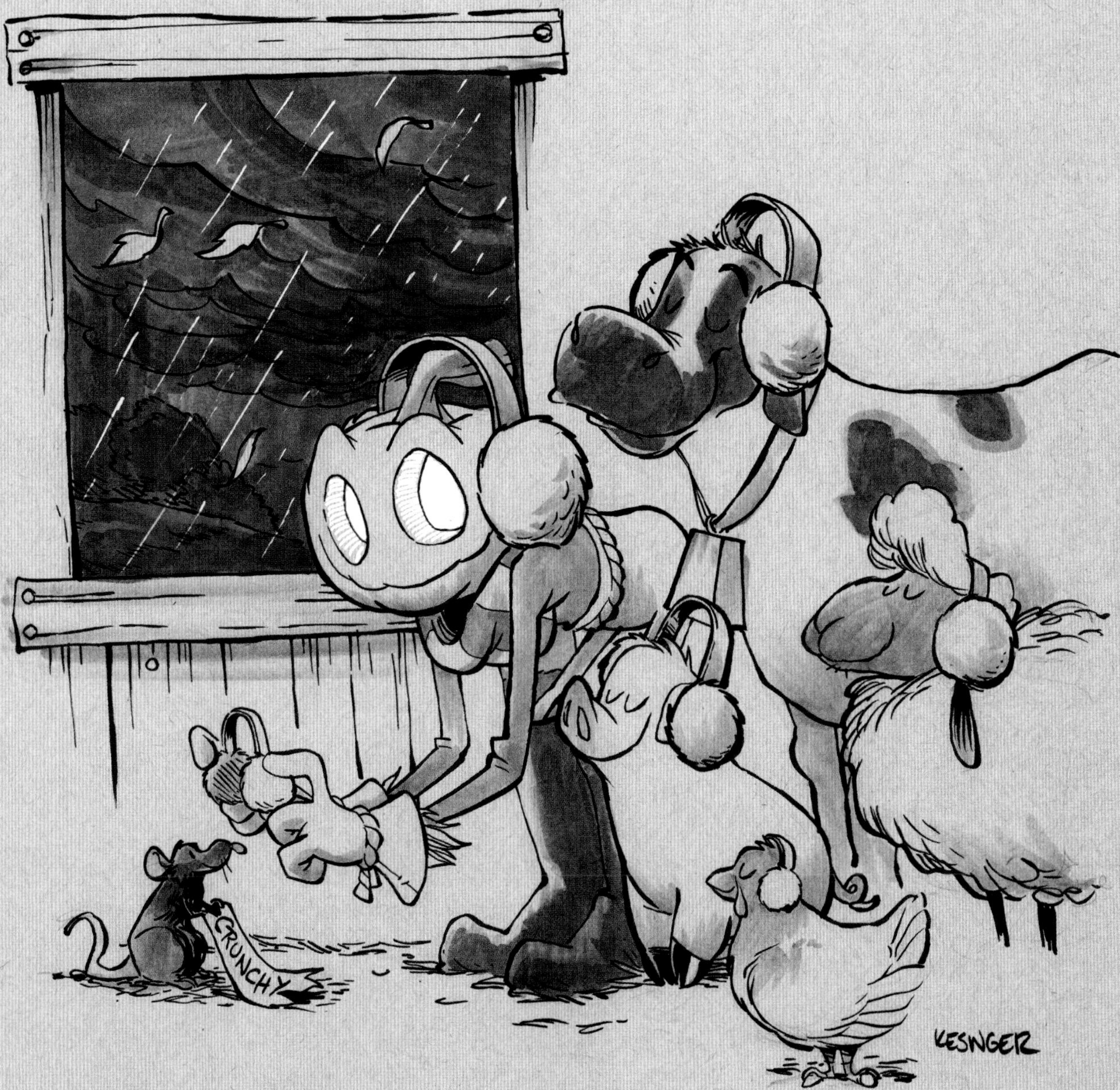
CRUNCHY
KESINGER

PATENT 1225:

"The Jolly-Wrapper"

This seasonal innovation is designed to alleviate the stress that surrounds so much of the holidays.

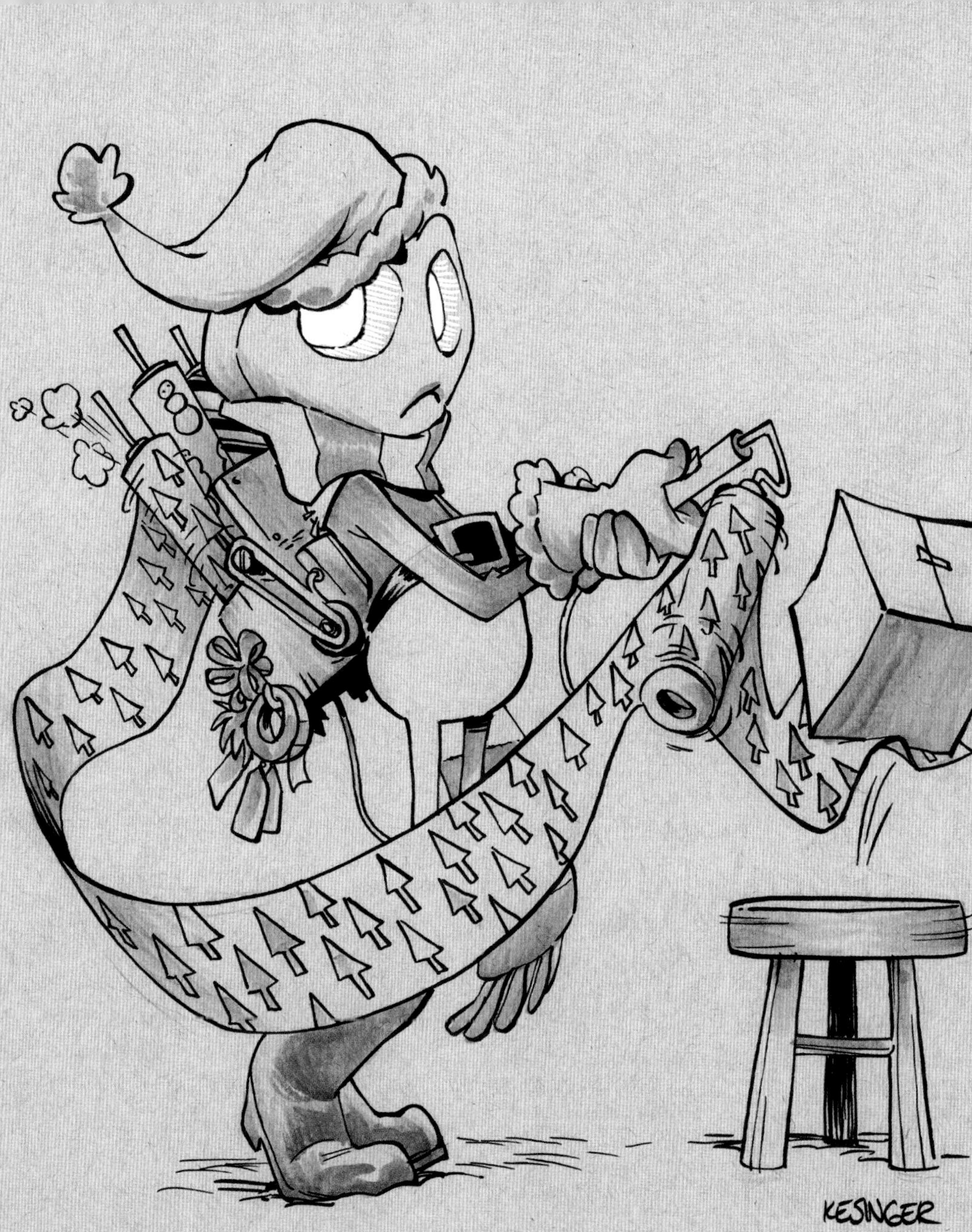
KESINGER

Protip:

Remember to step away from the workbench every now and then to enjoy new perspectives and partake in festive traditions.

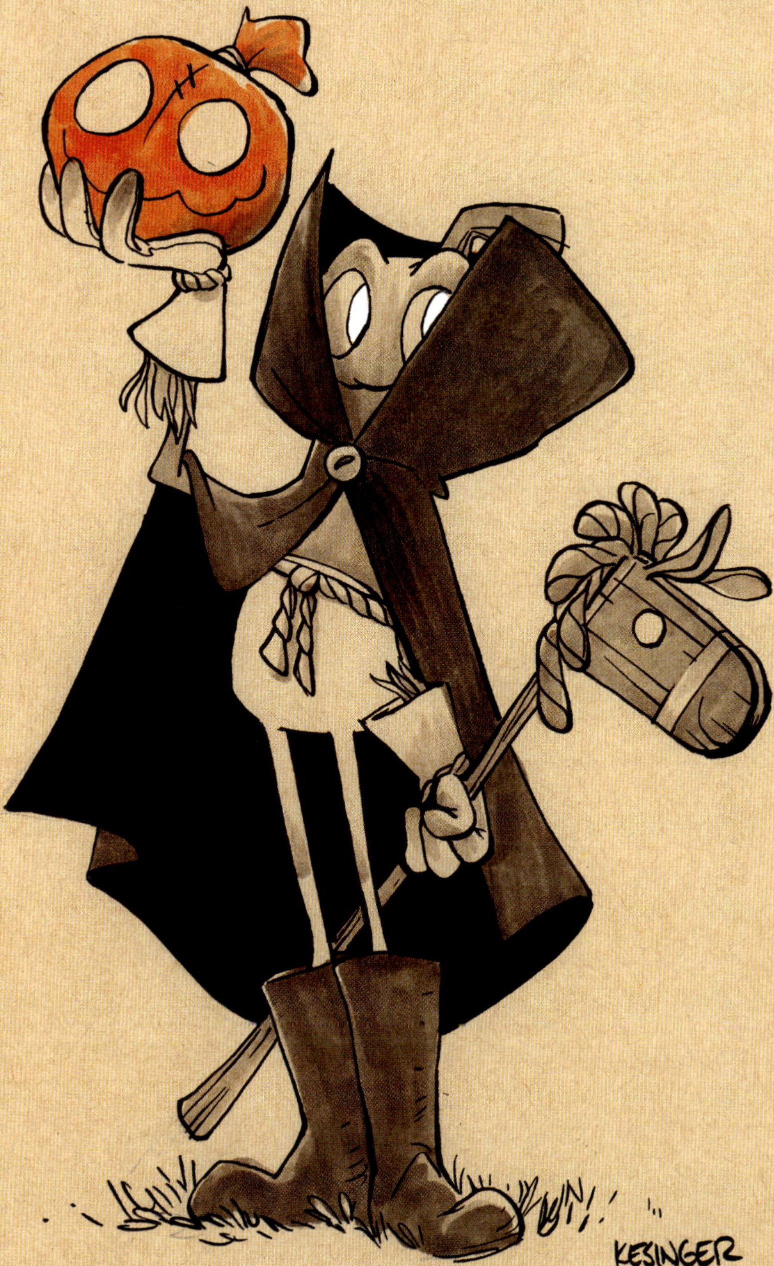
KESINGER

PATENT 140.55:

"The Bright-Bulb"

Engineered to project over 90,000 lumens, this lamp requires an enormous amount of energy and has yet to be tested.

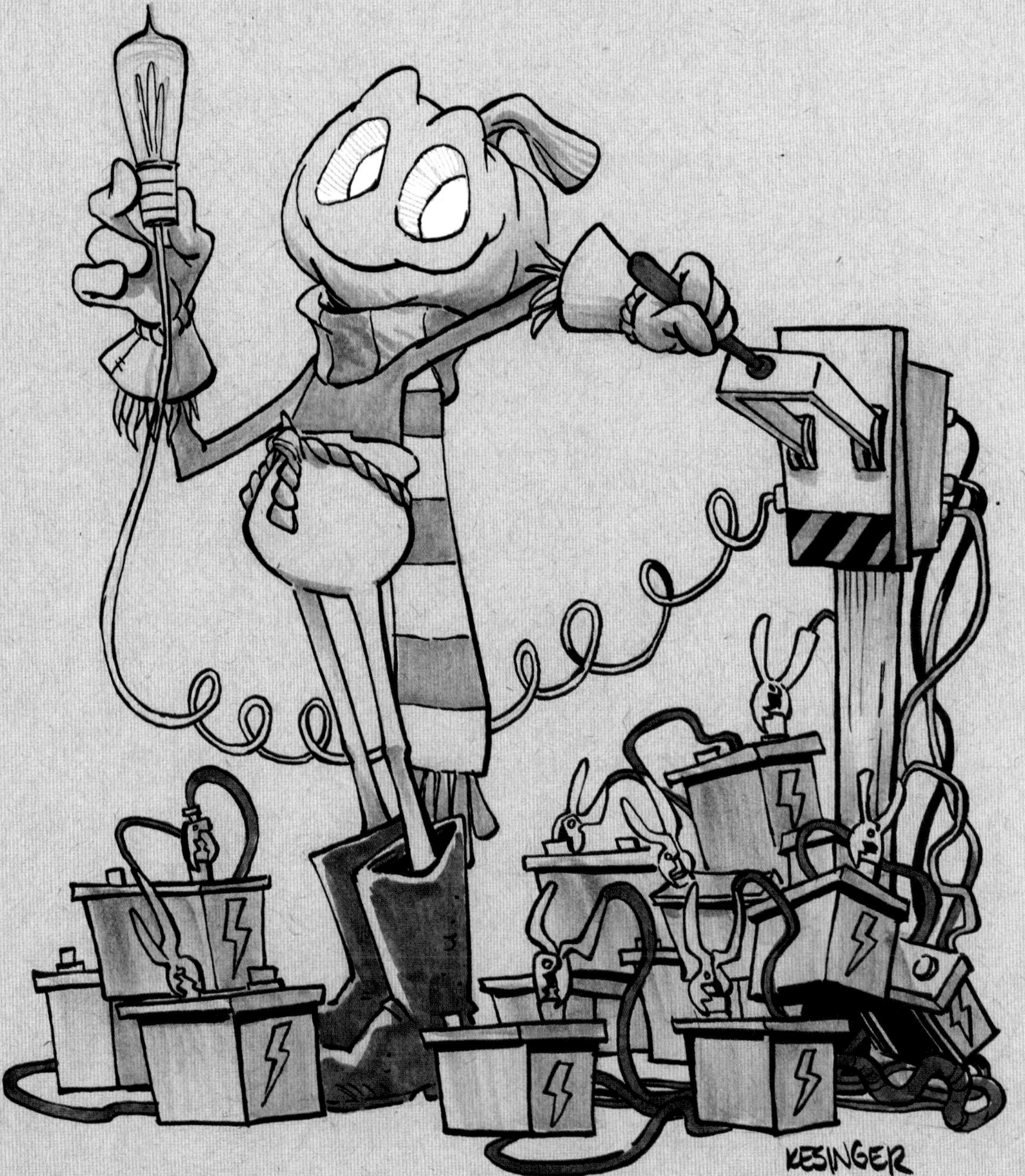
KESINGER

To-do list item #78:

Invent friends to celebrate life events with.

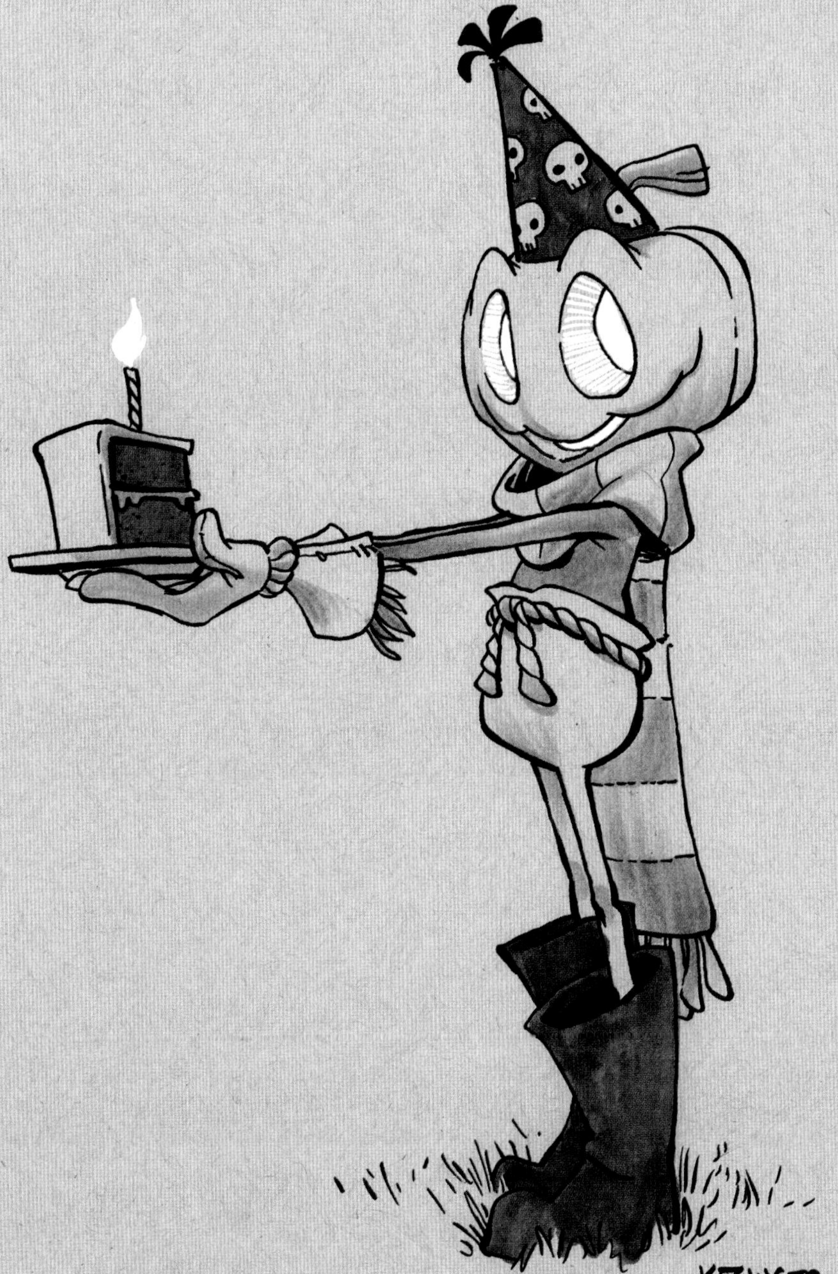
KESINGER

PATENT 333:

"The Grass-Guzzler"

Engineered to cut 0-60 acres of grass in under a minute, you are encouraged to buckle up before ignition.

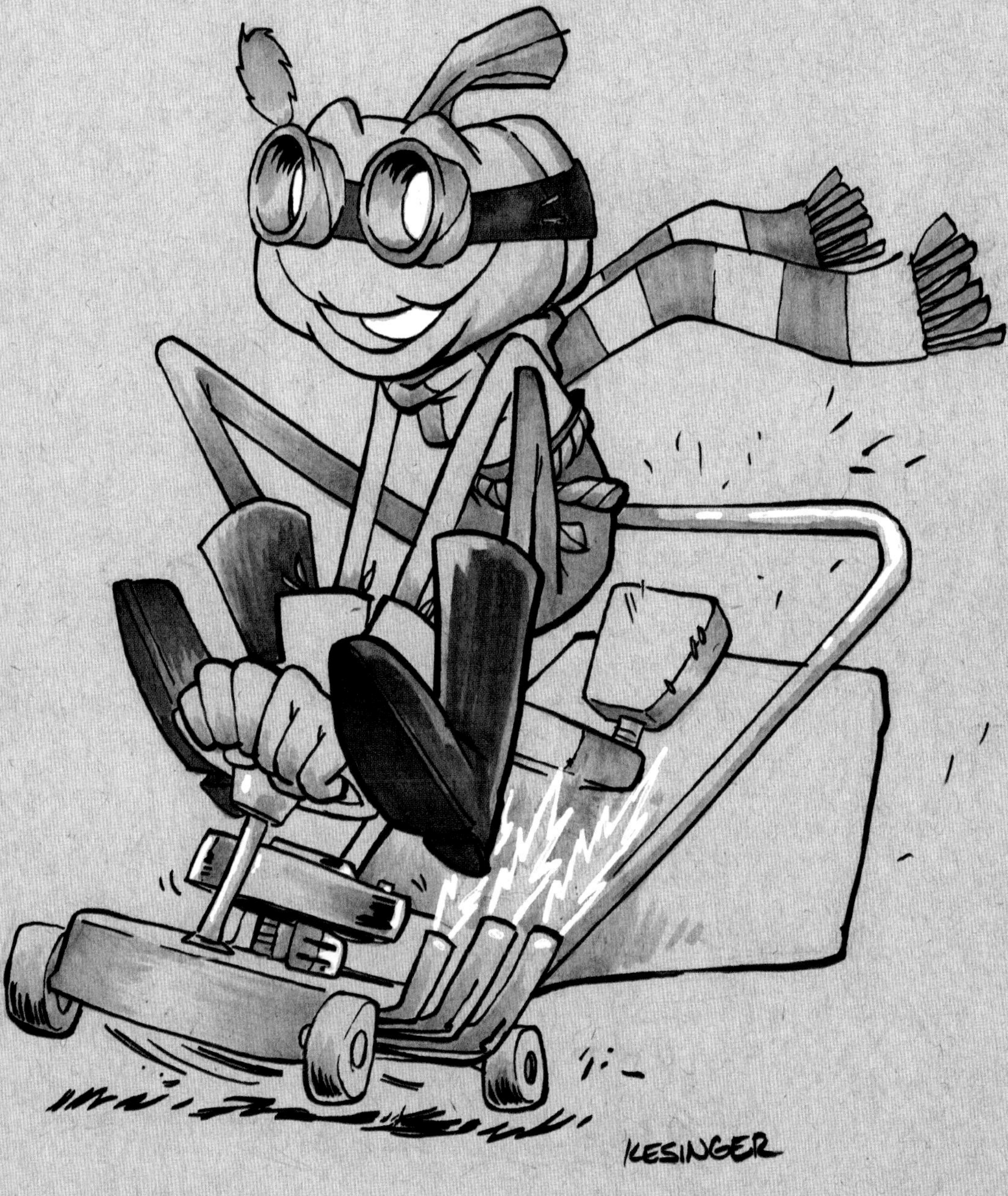
KESINGER

PATENT 56.0045:

"The Boot-Boosters"

Put a spring in your step and yield twice the crops at the same time!

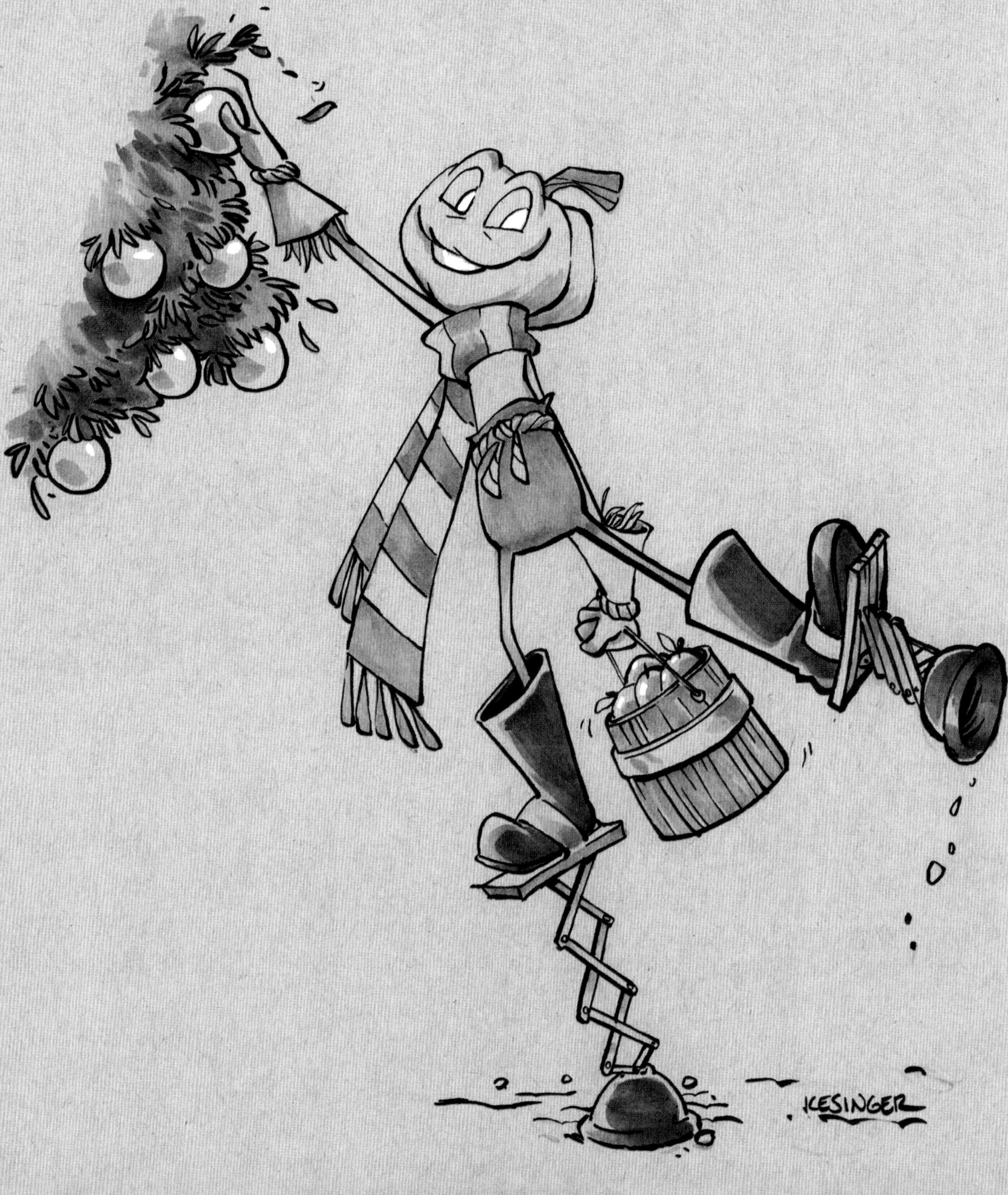
KESINGER

Protip:

What good is an invention if it can't be enjoyed by others?

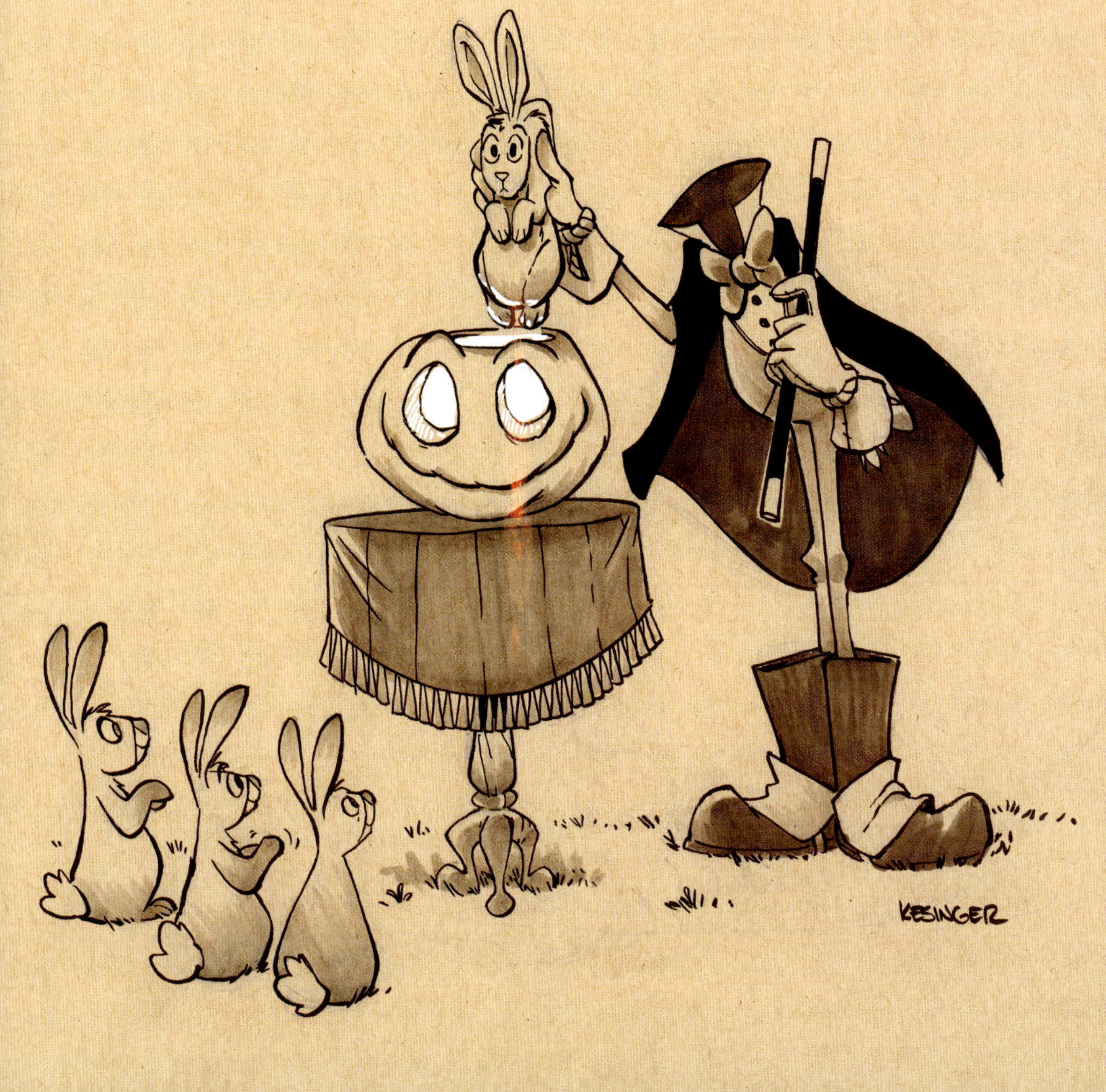
KESINGER

PATENT 9786-G:

"The Joy-Stick"

Not all inventions have a purpose.
This is one of those inventions.

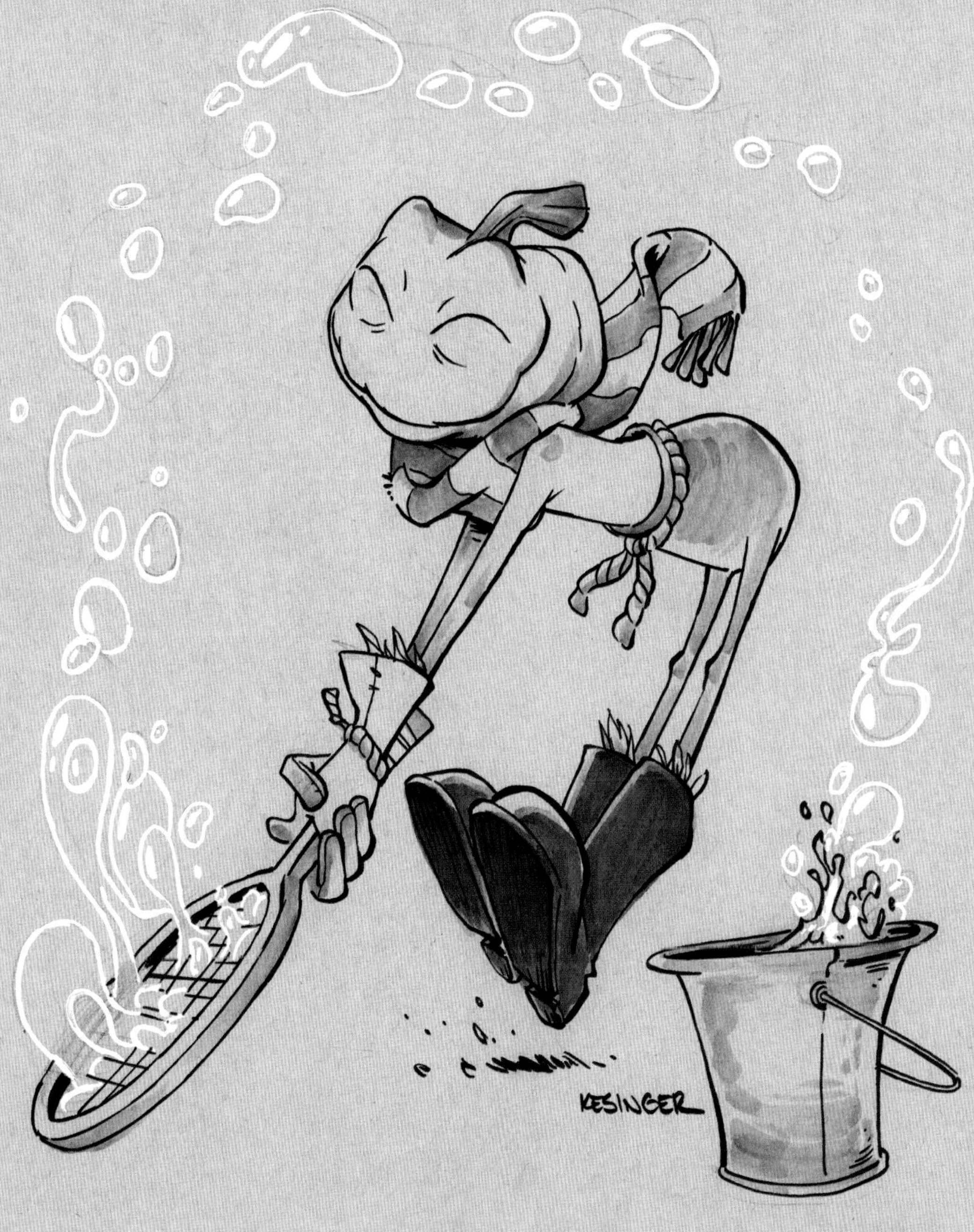
KESINGER

PATENT 1989LM:

"The Tall-Tail"

Fishing has never been easier with this perfectly camouflaged contraption.

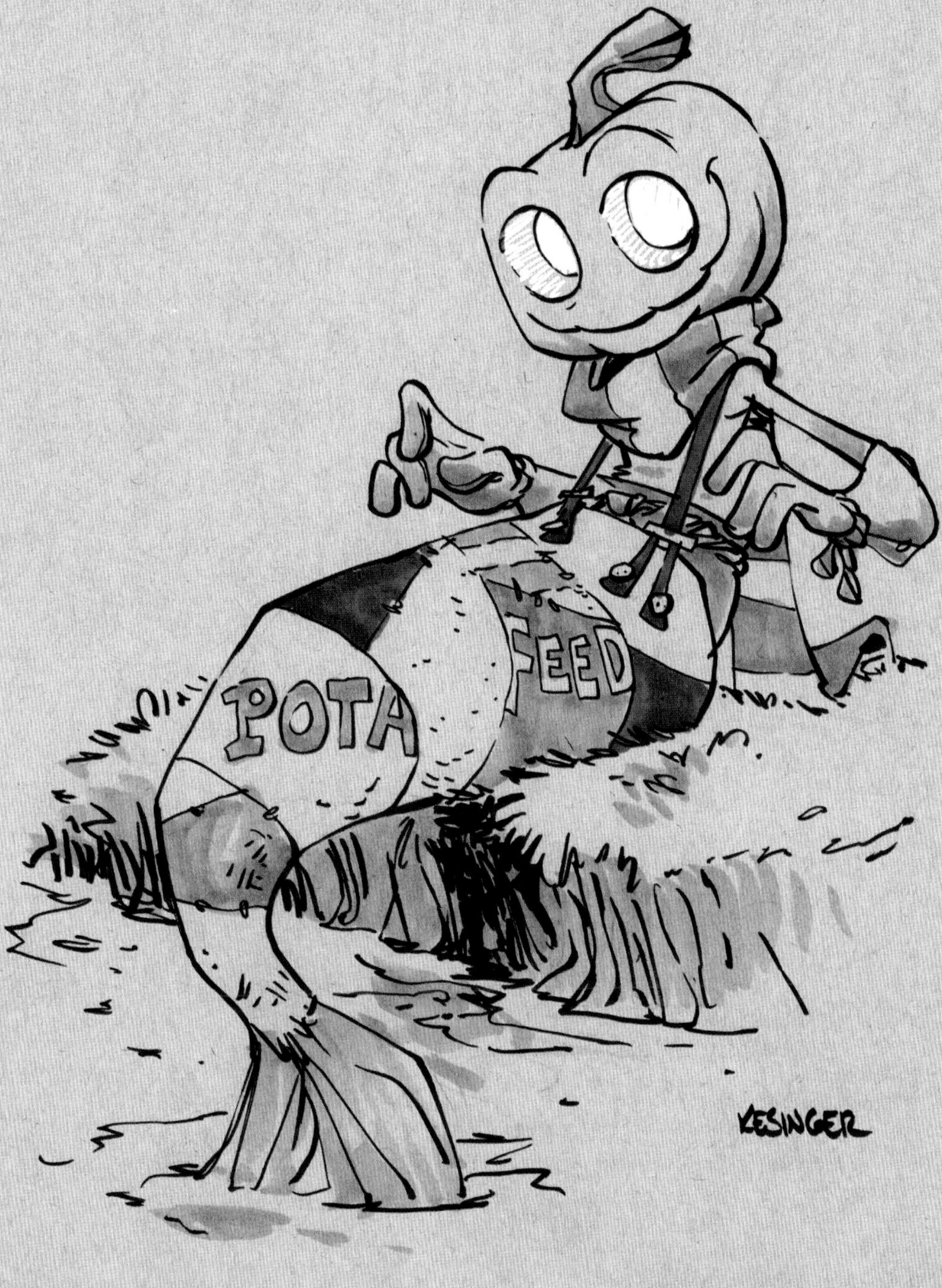
POTA
FEED
KESINGER

Protip:

Dreaming is doing. No one is ever an overnight success but with hard work and dedication your dreams will come true.

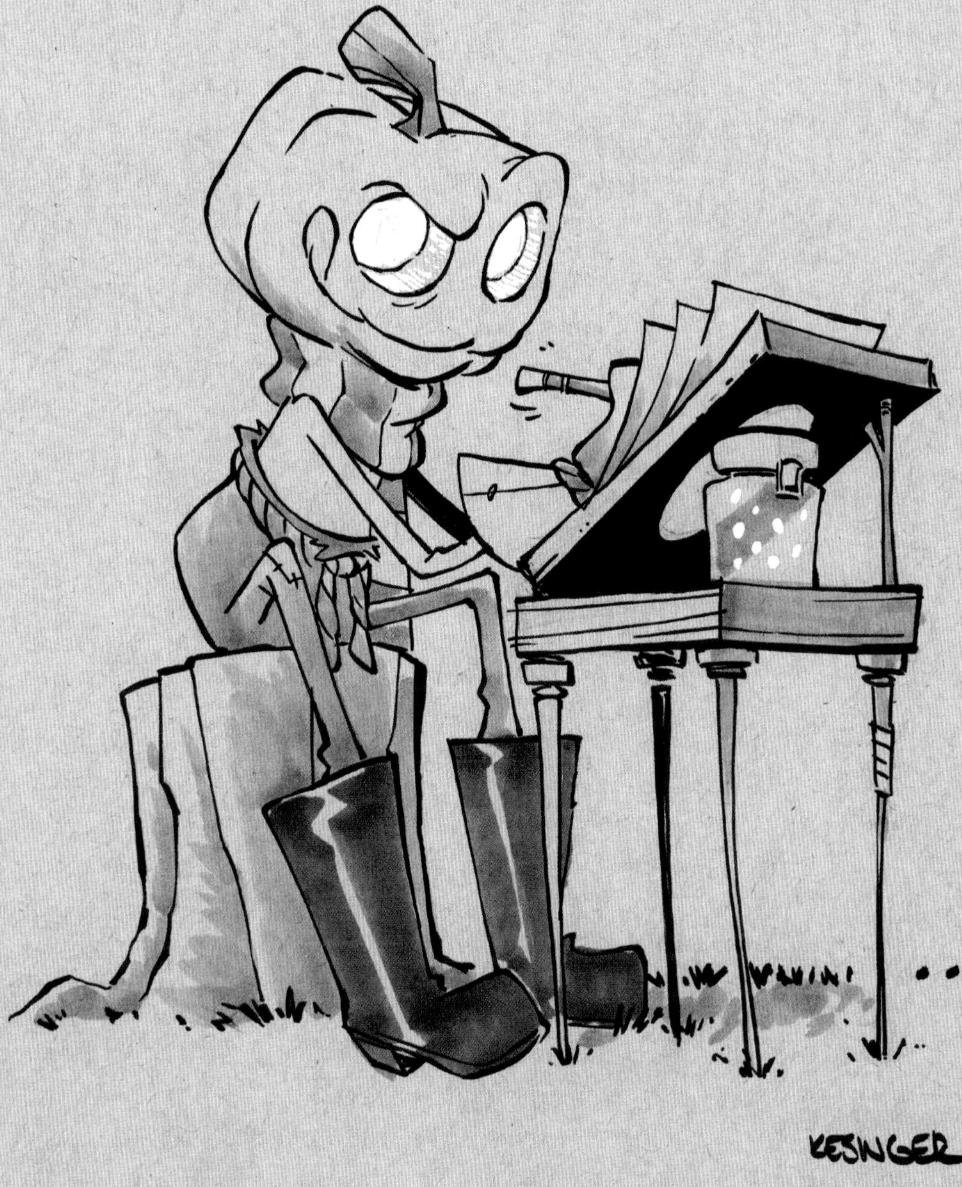
KESINGER

SCARECROW SUPPORTERS

#BAMO2016
A. Singer
A Sleepy Soul
Aaron Jamieson
Aaron Newton
Abhilash Sarhadi
Abigail Vodraska
Adam Shephard
Adrian Polo
A. J. Sabino
Alan Litteneker
Alan Smithee
Alex & Rachel Tiedemann
Alexa Montgomery
Alexia Charlesworth
Alexis Wallick
Alicia Eilers
Allan A. MacBain
Amanda Johnson
Amanda St. Germain
Amanda Weeks
Amandine
Amber Asmussen
Amy-Maree Langman
Andrew & Ivy
Andrew G. Lucas
Angie Reiber
Anne Brightman
Anne-Sophie Poblon
Anthony Ramirez
Anonymous
Ari Crane
Armond Netherly
Ashley Brinegar
Ashley Pearce
Athena & Desserä
Audrey Miles
Aureanus
Avery Belanger
Barry T. Drake
Baxter Taylor
Becky Ballantine
Becky Smith
Ben
Ben Mordue
Ben Norris
Bethany Hughes-Diem
Bobbie Kirby
Bre Bowker
Brett Reighn
Brian, Lauren, & Claire
Brianna Kelly
Caeodos
Caitlin M. Riley
Caliope
Calvin Newman
Candice Grundy
Carla Kaessinger Coupe
Carlos Melgar
Carrie Haughey
cat61986
CBG
Cdfisher09
Cecil Keebler
Charles Kuzmech
Charlie Blake
Charlie Rogers
Charlie Tucker
Chase Allen Strickland
Chelsea Olson
Cherie Luras
Cheryl K. M.
Chris "Running" Lin
Chris Pilla
Chris Roberson
Chris Smith
Christina R.
Christopher Medarac
Claire & Emma Gray
codingCookie
Colin Jordan Taylor
Colleen Greenlee
Colleen River Murphy
Connor Thomas
Corinna Cornett
Corky LaVallee
D. & R. Zukley
D. V. Diringshofen
Dan Weston
Daniel G. Pittenger
Daniel McManus
Daunell Jean Barton
Dave Derrick
David Awesome Cole
David denBoer
David Wills
Derek J. Bush
Don Tardiff
Donald Ogg
Donzie Mathis
Doug Walker
Dr. Joy Trujillo
Dr. Kate & Friends
E. R. Auld
Eagle Archambeault
Elaine Dailey
Elaine M. Cassell
Elizabeth Leiby
Elizabeth Thrash
Ellen Hubbard
Eloise Nolte
Emily Collins
Emma Luigard
Eric Renn
Erin Sayers
Ethan R. Brown
Evan and Shervin Farivar
Everett A. Warren
Fion J. Lau
Fix Hawksby
FoDeJo
Fox Lundgren
Francis de Guzman
Fred W. Johnson
Gabriel Glachant
Gene Branaman
Genevieve Frances Biro
George Gaspar
Gershowitz Family
Gil Pedersen
GoNXaS
Greg Tangey
guardian__J
Gwyn Brown
Hannah England
Hannah Fraley
Heart of Oak
Heather Bowland
Henrik Kristensson
Honey B
Ivy
J. D. Kelleher
J. Rizzo
Jack Cummings
Jack Lope Vollman
James Crider
James R. Crowley
Jameson Couch
Jamie Thompson
Jan Faborsky
Jane Frances Schafer
Jane Young
Janette Moehnacz
Janine Brauer
Jase Membrey
Jasmine B.
Jasmine G. Lopez
Jason
Jason B. Milligan
Jason Whatever

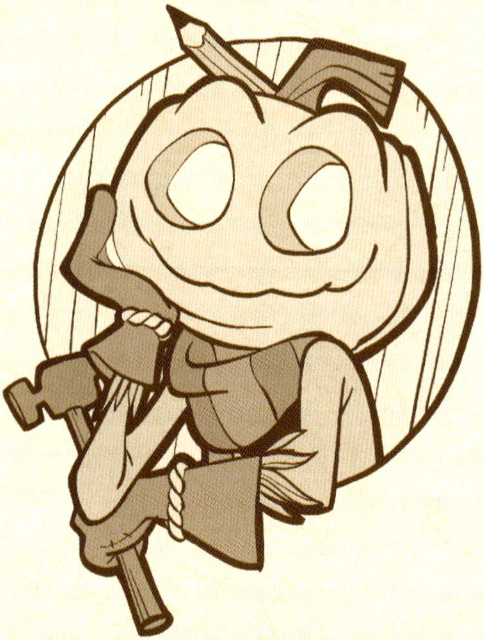

Jasper Allen-Kingsland
Jay N. Davis
J. Chism Studios
Jeff Lewis
Jeff Parsons
Jeffrey & Lisa Vaca
Jen- Brown
Jeni Danes
Jennifer Battan
Jennifer D. Andrews
Jennifer Hjelm
Jennifer K. Ryan
Jennifer Sutton
Jess Hogan
Jessica Alesna
Jessie Hippert
Jim Crawford
Jimmy "Sotemn" Hsieh
JimRomer.com
Jinx
Jinxe
Jo & Kate Toms
Jody L. Ransome
Joe Weaver
JoeCat
John MacLeod
John P. Bengier
John Semenec
John Ver Linden
Jonathan D Phillips
Jordan Jelmeland
Joseph "Launchpad" Carney
Joshua Gillman
Juan Loayza
Julia Nielsen
Juliana V.
Julia Whitaker
Julie & Ian Greig
Julie Gullman
Julie Mikesell
Justine Bergman
Karl Meister
Kat Wallis
Katarina
Kate Moon
Katherine S
Kathryn Berghold-Moheit
Katie Fabrick

Keeley Pollock
Keena Finney
Kelly Butler
Kelsey B.
Kendra S. Skinner
Kendra Witherspoon
Kevin Deters
Kevin H.
Kimmi Allbee
Kinderknecht
Kirsty Kilpatrick
kre luvs C&H/SW
Kristen Hariton
Kylie N.
Lauren H.
Lee Millar
Leila Tilghman
Liam
Linda Smit Poche
Lisa D. Fischer
Lisa Hutcheson
Lisa Labrunda
Lisa Sustaita
Liz Cromwell
Lucia Soltis
Luke Flowers Creative
Maggie Allen
Maggie Jan
Mal Souza
Mamie Scholle
Mandy Hobson
Mandy Nadeau
Marek Bláha
Maria João Castelão
Marisa Faiz
Mark Mitchell
Mark Thelosen
Marne E. Gifford
Marty Trengove
Mary Ann Shuman
Mary Francis Williams
Mary M. Crocker
Mathias Wong
Matt Acosta
Matt Subrizi
Mayer Brenner
McKenna R. Derr
Meaghan Karl

Megan Real
Mélisa White
Mia Anderson
Michael Acovio
Michael Hahn
Michele Hall
Michelle Deibner
Mike Kunkel
Miss Eliza
Moira Manson
Monica A. Rocha
Monica Anderson
Monica Bruenjes
Morgan Witkowski
Moria Trent
Mr. Josh Mistretta
Mr. Yetti
Nate Little
Nick & Connie
Nicole M. Tucker
Nupoor Kalke
Nuria Flores
Oldhippiedragon
Pabkins T. Jensen
Paige Leach
Patty Fotheringham
Paul (Old Buddhist Crab)
Thorgrimson
Paul M. Reynolds
Peter Pakovics
Phineas Merriman
Pia Conrad Mortensen
PoppyPhlox
PrescriptionForAdventure
R. J. H.
Rachel Jacobson
Razael Fletcher
R. D. Rocha
Riley Wesley Tannler
Ron Pack
S
S. Michener
Samantha Squires
Sara Glassman
Sara Street
Sarah C. Orkwis
Sarah Drake
Sarah Hoover

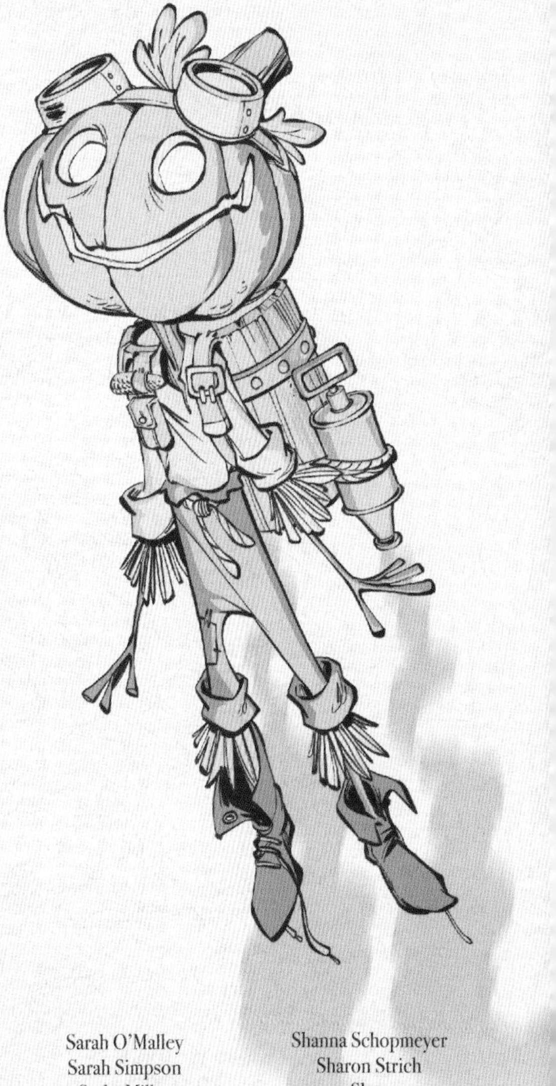

Sarah O'Malley
Sarah Simpson
Sasha Miller
Schkan
Scott Elson
Sean Patrick Terska
Sebastian Dess
Shadow
Shaken

Shanna Schopmeyer
Sharon Strich
Shay
Skyline Music
SmallRiniLady
Sonja Harris
Sonia Lai
Sophia Dittman
Squid Kids Ink

Stephanie
Stephanie D. Anderson
Stephanie Smith
Steve Muench
Steven Ng
stormibear59 aka Natasha
Anne Lux
Suzy Creamcheese
Swazzle
T. Smith
The Alvis Family
The Anonymous Backer
The Atkinson Family
The Bembers
The Bonines
The Bush Family
The Campbell Clan
The Celaya Family
The Crevani Family
The Elliott Family
The Elliotts
The Fitches
The Garoutte Family
The Gelinas Family
The Geoghegan Family
The Herzberg Family
The Hylands
The Karteczkas
The Loung Family
The MacIntyre Family
The Malinoski Family
The Melrose Family
The Moffatt Family
The Moore Family
The Padilla Family
The Parks
The Q's
The Quinn Girls
The Sanchez-Cruz Family
The Schoenthal Family
The Sinay Family
The Spenards
The Three Blairs
The Yangs
Theda LaStrel
Thomas Millesich
Tim Harwood
Tim Jordan
Timothy Stephens
Tinna Guðbjartsdóttir
Tom Deal - Deal With It
Tracey Fritz
Victoria G. Campbell
Vipi Ink
Wally Kendall
Wanda Aasen
Whitney Jelmeland
William McKinney
William Noffsinger
Wilson B. Thomas
Winston Kou
Yue Li
Zaiga K. Adamou
Zeke Hanson
Zoe Valenza

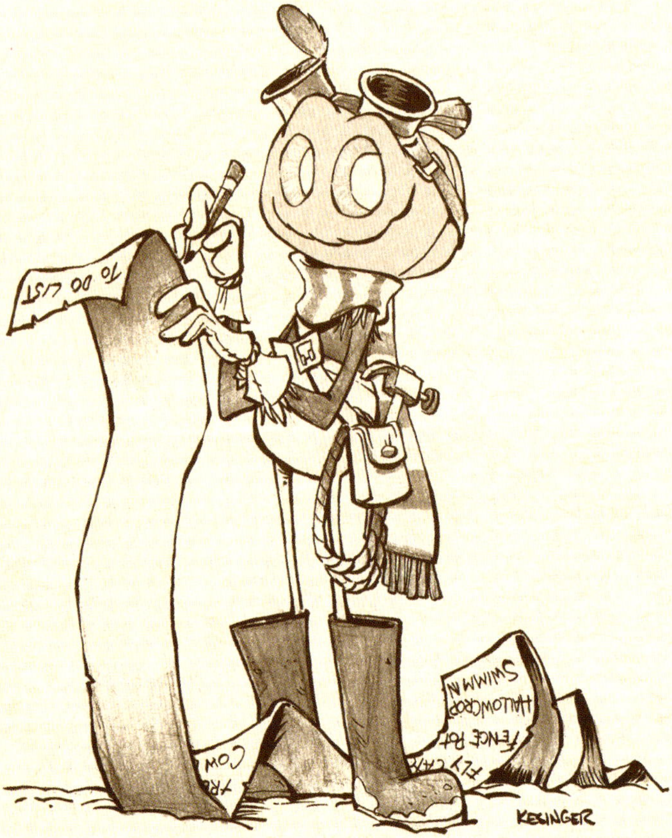

Also available from Brian Kesinger and Baby Tattoo Books®

BRIAN KESINGER'S PENNED DRAGONS
ISBN 978-1-61404-021-7

BRIAN KESINGER'S INKED TAILS
ISBN 978-1-61404-022-4

WALKING YOUR OCTOPUS
ISBN: 978-1-61404-006-4

TRAVELING WITH YOUR OCTOPUS
ISBN: 978-1-61404-009-5

TIME TRAVELING WITH YOUR OCTOPUS
ISBN: 978-1-61404-028-6

COLORING WITH YOUR OCTOPUS
ISBN: 978-1-61404-010-1

DRESSING YOUR OCTOPUS
ISBN 978-1-61404-016-3